TECHNIQUES AND METHODS OF RADIO-ASTRONOMIC RECEPTION

TEKHNIKA I METODY RADIO-ASTRONOMICHESKOGO PRIEMA

ТЕХНИКА И МЕТОДЫ РАДИО-АСТРОНОМИЧЕСКОГО ПРИЕМА

The Lebedev Physics Institute Series

Editors: Academicians D. V. Skobel'tsyn and N. G. Basov

P. N. Lebedev Physics Institute, Academy of Sciences of the USSR

Recent Volumes in this Series

Proceedings (Trudy) of the P. N. Lebedev Physics Institute

Volume 93

Techniques and Methods of Radio-Astronomic Reception

Edited by

N. G. Basov

P. N. Lebedev Physics Institute
Academy of Sciences of the USSR
Moscow, USSR

Translated from Russian

SPRINGER SCIENCE+BUSINESS MEDIA, LLC

Library of Congress Cataloging in Publication Data

Main entry under title:

Techniques and methods of radio-astronomic reception.

(Proceedings (Trudy) of the P. N. Lebedev Physics Institute, Academy of Sciences of the USSR; v. 93)
Translation of Tekhnika i metody radio-astronomicheskogo priema.
Includes index(es)
1. Radio astronomy—Addresses, essays, lectures. 2. Radio telescope—Addresses, essays, lectures. I. Basov, Nikolai Gennadievich, 1922-   II. Series: Akademiia nauk SSSR. Fizicheskii institut. Proceedings; v. 93.
QC1.A4114 vol. 93   [QB478.5]   530′.08s [522 .682]
                                                                              78-26720

ISBN 978-1-4684-1640-4   ISBN 978-1-4684-1638-1 (eBook)
DOI 10.1007/978-1-4684-1638-1

The original Russian text was published by Nauka Press in Moscow in 1977 for the Academy of Sciences of the USSR as Volume 93 of the Proceedings of the P. N. Lebedev Physics Institute. This translation is published under an agreement with the Copyright Agency of the USSR (VAAP).

CONTENTS

AUTOMATION OF RADIO-ASTRONOMIC INVESTIGATIONS USING THE RT-22 RADIO TELESCOPE WITH AN ELECTRONIC COMPUTER

A. V. Kutsenko, B. A. Polos'yants, Yu. M. Polubesova, and R. L. Sorochenko

The basic requirements of a system for automating radio-astronomic investigations with the RT-22 radio telescope are formulated, and its functional arrangement is described. The reasons are given for the selection of the system's programming and configuration, and the principal operational algorithms are presented.

The 22-meter RT-22 radio telescope of the P. N. Lebedev Institute of Physics [1] has continued, up to the present time, to be one of the largest radio telescopes in the world that works in the millimeter wave band. This radio telescope is equipped with highly sensitive apparatus that utilizes quantum paramagnetic amplifiers (masers). The current level of radio-astronomic research requires that the instrument system of a radio telescope include an electronic computer as an essential component, in order to ensure the control of all the processes of observation. Below is described the design for the total automation of radio-astronomic observations using the RT-22 radio-telescope with an M-6000 electronic computer.

1. The Requirements of the System and Its Functional Arrangement

The system for automating the RT-22 involves the following operations: (1) observation programming in accordance with input data; (2) aiming and guidance of the radio telescope in accordance with the program; (3) multichannel collection of information and its storage and preliminary processing; (4) programmed control of the radio-astronomic receiving apparatus, and calibration and verification of the necessary parameters; (5) recording of data on the radio telescope's position and on the signals being received, and their output in the form of quick information for observation and in a form permitting further processing with large electronic computers or with the radio telescope's own computer during time not used for observations.

The operation of the entire system must proceed on a real time scale with the precise correlation needed for astronomic measurements. A diagram of the automation system for radio-astronomic observations, constructed on the basis of the above-listed requirements, is presented in Fig. 1.

1.1. Aiming and Guidance of the Radio Telescope

The aiming of the radio telescope at an assigned spot on the celestial sphere is one of the basic technical problems in radio-astronomic observations. With the RT-22's directional-

1

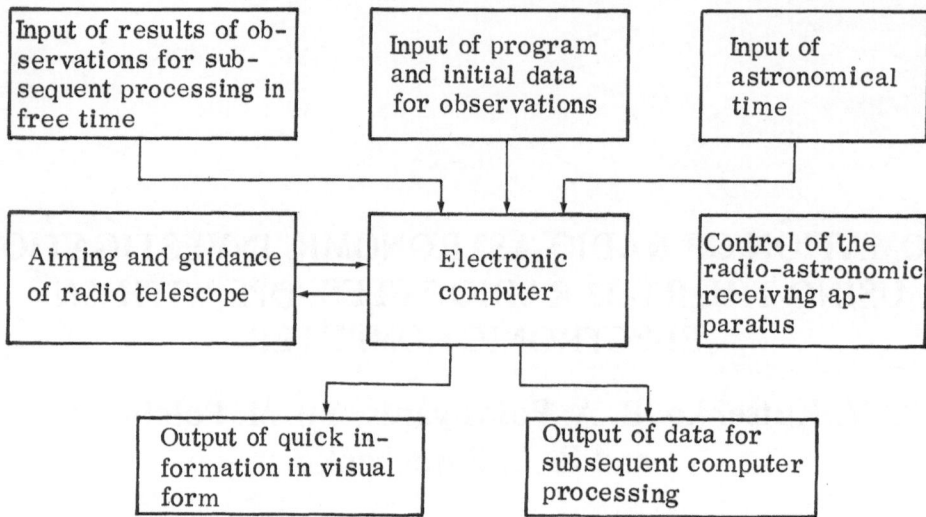

Fig. 1. Diagram of the automation of radio-astronomic observations with
the RT-22 radio telescope.

diagram width of 1.9 minutes of arc (at the shortest operating wavelength of 8 mm), the required spatial aiming accuracy should be not worse than 10"–15". A generalized diagram of
the radio telescope's guidance [2] along one coordinate is presented in Fig. 2. The computer,
operating on a real time scale provided by an outside timer, computes in steps of 0.5, with
coordinates and velocity along both axes. With the prescribed overall aiming error of 10" to
15", the required accuracy for computing the coordinates should be ~1", which corresponds
to 21 binary orders in azimuth and 19 binary orders in angle of elevation. The true position
of the radio telescope is determined by 18-order angle-code (angle-digital) data units installed
on the radio telescope's axes. The discrimination of the angular-mismatch or angular-error
signals and the display of the necessary quantities are done by the electronic computer.

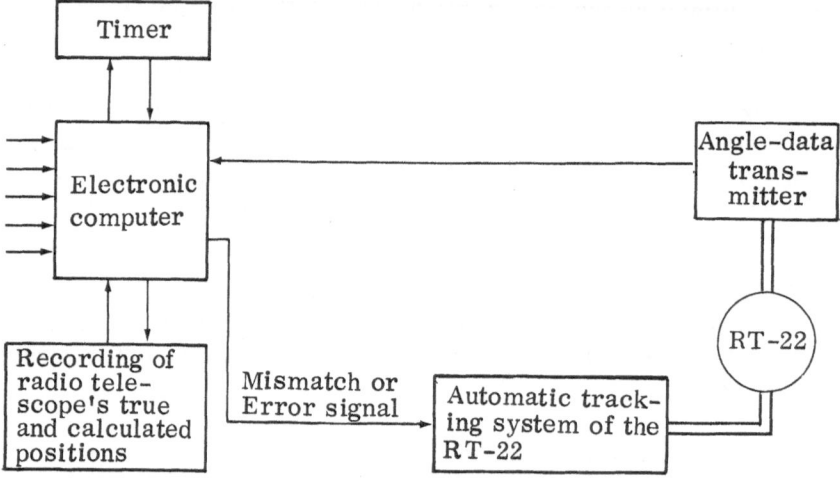

Fig. 2. Generalized diagram of the RT-22's guidance by the electronic computer, along one coordinate.

Depending on the program of scientific observations, the radio telescope's guidance system must secure:

1. The aiming of the radio telescope at the necessary point on the celestial sphere.
2. The tracking of the assigned point, which moves across the celestial sphere in accordance with the laws of motion of the stars, sun, solar-system planets, and moon. It is possible to follow two basic procedures in this program: (a) tracking by the aiming−withdrawal method, in which the radial telescope observes the source and then withdraws to a reference point, and (b) tracking, by the aiming−aiming method, in which the radio telescope, operating with a symmetrical diagram modulation, observes the source alternately in one or the other possible directions of reception [3].
3. The passage of the radio telescope's directional diagram through the selected point with a given velocity, that is, a scanning regime.

The aiming should be done both by the geometric axis of the radio telescope and by an axis offset from the geometric one through an angle of 1° in both coordinates.

The Radio Telescope's Guidance Algorithm. For computation of the radio telescope's position at the time of observations by the "aiming" and "tracking" programs, the initial data are: (1) the right ascension α of the point being observed; (2) corrections to the right ascensions for planets, sun, and moon, daily and hourly accordingly, $\Delta\alpha_m$ (mean); (3) declination δ; (4) corrections to the declinations for planets, sun, and moon, $\Delta\delta_m$; (5) universal stellar time S_0; (6) collimation correction of the radio telescope in azimuth, ΔA_c; (7) collimation correction of the radio telescope in zenith distance, ΔZ_c.

At the time of observations by the aiming−withdrawal program, to the initial data are added: (8) the coordinates of a reference point, α_2, δ_2; (9) the tracking times for the experimental and reference points, T_1 and T_2.

At the time of observations by the aiming−aiming method, to the initial data are added: (8) the difference between the two directions of reception in azimuthal coordinates (with arbitrary direction of the radio telescope to the zenith), ΔA_d, ΔZ_d; (9) the duration of observations in each of the directions of reception, T_1 and T_2.

At the time of observations by the scanning program, the scanning parameters are also initial data. Four kinds of scanning are possible.

Scanning along the Right Ascension α. Specified are: (1) right-ascension scanning range $\Delta\alpha_{lim}$; (2) scanning rate $\dot{\alpha}$; (3) declination scan interval $\Delta\delta_{sc}$; (4) declination scan limit $\Delta\delta_{lim}$.

Scanning along the Declination δ. Specified are: (1) declination scanning range $\Delta\delta_{lim}$; (2) scanning rate $\dot{\delta}$; (3) right-ascension scan interval $\Delta\alpha_{sc}$; (4) right-ascension scan limit $\Delta\alpha_{lim}$.

Scanning along the Azimuth A. Specified are: (1) azimuth scanning range ΔA_{lim}; (2) scanning rate $\dot{A}$; (3) zenith scan interval ΔZ_{sc}; (4) zenith scan limit ΔZ_{lim}.

Scanning along the Zenith Distance Z. Specified are: (1) zenith-distance scanning range ΔZ; (2) scanning rate $\dot{Z}$; (3) azimuth scan interval ΔA_{sc}; (4) scan limit ΔA_{lim}.

In addition to the above-mentioned initial data, the various programs for the computer solution of the guidance problems involve the following:

Absolute constants: (1) the radio telescope's geographic latitude φ; (2) the radio telescope's geographic longitude λ; (3) the time-zone number N; (4) the refraction constant (see below);

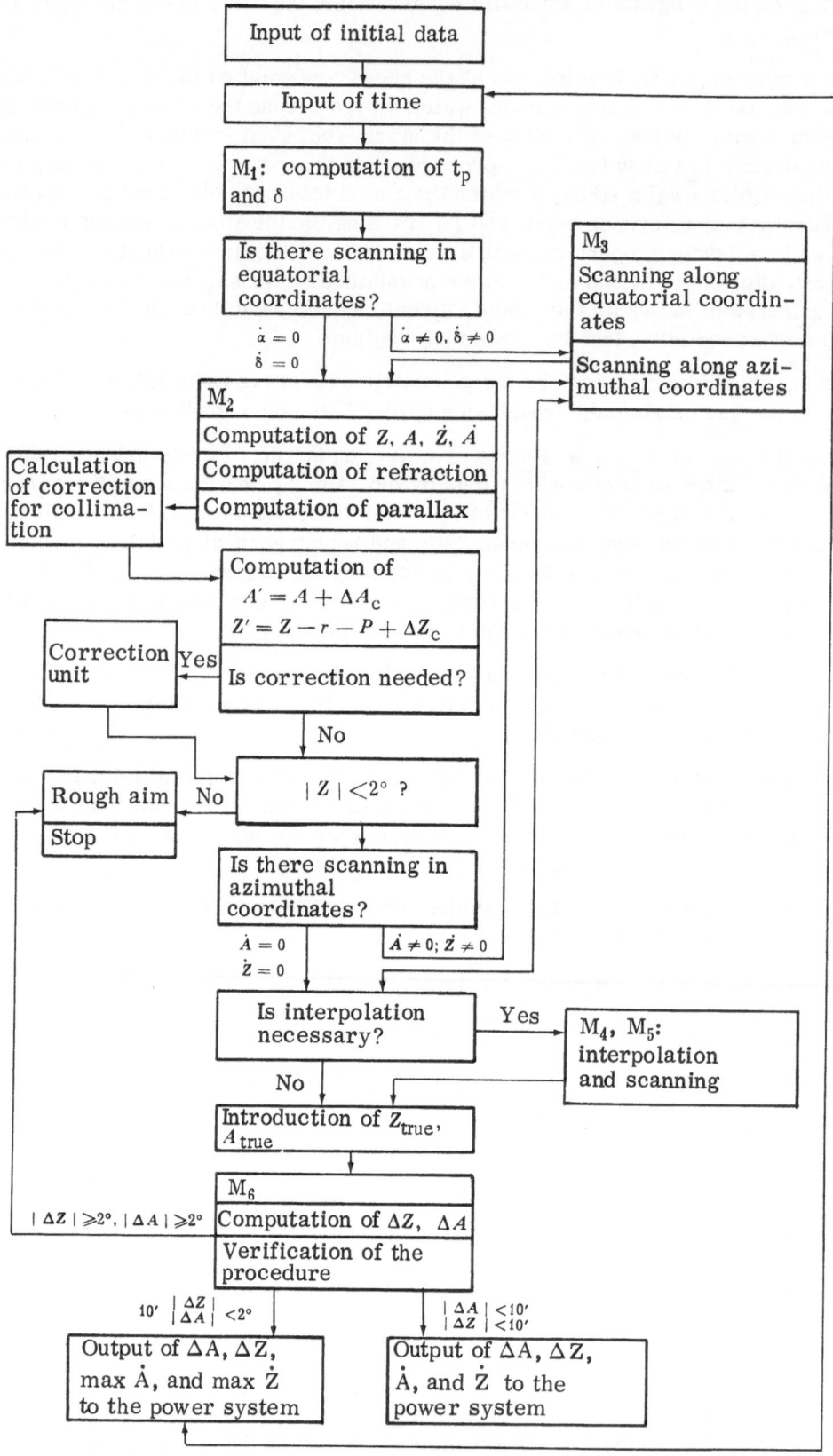

Fig. 3. Block diagram of the radio telescope's guidance algorithm.

Constants at the time of the observations: (1) the environmental temperature T; (2) the pressure P; (3) the spherical constants for observation of the moon.

The radio telescope's guidance algorithm, a block diagram of which is presented in Fig. 3, can be described by means of a series of operators subsequently designated as M_1, M_2, M_3, ...

Operator M_1 performs the computation of the hour angle t_m and declination δ of the experimental point at any time. This operator consists of three basic equations:

$$t_p = S_0 - \alpha + \lambda + [1 + {}^1\!/_{24}\ (\Delta t_\gamma - l\Delta\alpha_m)]\ (T_d - N - 1 + \Delta t) + \Delta\alpha_m T_{emn}, \tag{1}$$

where T_d is the local or district time; N is the number of the time zone; Δt is the computer's problem-solving time; $\Delta\alpha_m$ is the hourly mean correction for the moon and the daily mean correction for the planets and the sun (from the almanac); T_{emn} is the ephemeris time for the moon corresponding to the moment of initiation of lunar tracking (from the almanac); l = 24 for the moon and l = 1 for all the other celestial bodies; Δt_γ is the coefficient of conversion of mean time to stellar time;

$$\delta = \delta_0 + \Delta\delta, \tag{2}$$

where $\Delta\delta = (\Delta\delta_m/24)/(T_d - N - 1)$ for all objects except the moon; δ_0 is the declination of the source: $\Delta\delta_m$ is the mean correction to δ (from the almanac) (for stars it is zero); for the moon

$$\delta_{mn} = \delta_0 + \Delta\delta_m(T_d - T_{dmn} + \Delta t), \tag{3}$$

where δ_0 is the declination of the moon; $\Delta\delta_m$ is the daily mean correction to δ (from the almanac); $T_{dmn} = T_{emn} + N + 1$.

Operator M_2 determines the position and the rate of the radio telescope in the azimuthal system at a given hour angle t and declination δ:

$$\cos Z = \sin\varphi\cdot\sin\delta + \cos\varphi\cdot\cos\delta\cdot\cos t_p, \quad \sin Z = \sqrt{1 - \cos^2 Z},$$
$$\cos A = \frac{\sin\varphi\cdot\cos Z - \sin\delta}{\cos\varphi\cdot\sin Z}, \quad \sin A = \frac{\cos\delta\cdot\sin t_p}{\sin Z},$$
$$\dot{Z} = \cos\varphi\cdot\sin A, \quad \dot{A} = \sin\varphi + \cos\varphi\cdot\cos A\cdot\cot Z.$$

Corrections for the refraction r and for the source's parallax P are introduced into the computation of Z.

The parallax is computed from the formula $P = P_p \sin Z$, where P_p is the horizontal parallax of the point-source (from the almanac). Thus, the definitive value of the zenith angle is

$$Z' = Z + P + r.$$

The RT-22 radio telescope uses a system of antenna excitation with exciters displaced from the center of the reflector, in the horizontal plane [3]. As a result, the electrical axis differs from the geometrical axis: There is a collimation error A_c which must be taken into account in accordance with the equation $\Delta A_c = \Delta\varphi/\sin Z$. For different wavelengths, under the conditions of the RT-22, $\Delta\varphi$ is different. For example, for the 8 mm wavelength, $\Delta\varphi$ = 11.4'. Finally, we have

$$A' = A + \Delta A_c.$$

Operator M_3 provides for the scanning program:

$$\delta_a = a\Delta t, \ \Delta a_i = \Delta a_{i-1} + \delta_a, \ a_c = a + \Delta a_i, \quad \Delta b_i = \Delta b_{i-1} + \Delta b_c,$$
$$b_c = b + \Delta b_i,$$

where a is the smoothly varying coordinate, b is the coordinate that changes discontinuously, a and b are scanning parameters before a scan, a_c and b_c are the scanning parameters after the scan, Δa_{i-1}, Δa_i, Δb_{i-1}, Δb_i are two consecutive values of the scanning corrections, Δb_c is the scan step or interval, and Δa_{lim} and Δb_{lim} are the limiting values of the scanning. The parameters a and b can be either pair of coordinates: α and δ or A and Z, and in each pair the places can be changed.

For enhancing the reliability of the radio telescope's tracking system, operators M_4 and M_5 are introduced into the computer-control program.

Operator M_4 carries out linear interpolation. The motion of the object being investigated, across the celestial sphere, is described as a smooth function. In this case

$$A\ (t_1) \simeq A\ (t_0) + \dot{A}\ (t_0)\ (t_1 - t_0),$$
$$Z\ (t_1) \simeq Z\ (t_0) + \dot{Z}\ (t_0)\ (t_1 - t_0),$$

where $(t_1 - t_0)$ is the time interval between two cycles or series of computations, and $A(t_0)$, $Z(t_0)$, $\dot{A}(t_0)$, and $\dot{Z}(t_0)$ are values of the coordinates and velocities of the previous cycle.

The A and Z thus obtained are interpolated values.

Operator M_5 compares the computed values of the coordinates with the interpolated values.

If the difference between the velocities exceeds a prescribed limit, then the telescope's tracking drive mechanism is given control signals of the preceding cycle. If the computed values of the coordinates differ from the interpolated values by more than 2 seconds of arc (with a time interval between cycles of 0.5 sec and an acceleration of 1 second of arc per second), then no changes in the calculations are made. Otherwise, the computed values of the coordinates are replaced by the interpolated values.

In the case of the system being given interpolated values three cycles in succession, the computer is disconnected. This means that there has been a failure either in the computer or in the data units.

Operator M_6 carries out the calculation and the delivery of the control signals:

$$\Delta Z = Z_{true} - Z_{calc}, \quad \Delta A = A_{true} - A_{calc},$$

where Z_{calc} and A_{calc} are the calculated values of the coordinates, and Z_{true} and A_{true} are the true values taken from the data units.

Also carried out is the verification of several conditions that are necessary for the system's operation: If ΔA or ΔZ is greater than 2°, then a "rough-aim" signal is delivered; the aiming of the radio telescope at the object to be studied is carried out if the amount of mismatch does not exceed 2°; if the amount of mismatch does not exceed 10', then the tracking procedure is carried out.

In addition to the above-mentioned six operators, the control-problem algorithm involves also a correction or adjustment unit. The necessity of its use is due to the fact that, as a result of the radio telescope's inaccuracy (its horizontal axis is not strictly horizontal; its vertical axis is not strictly vertical; there may be a twisting of the axes, a deformation of the paraboloid, etc.), the actual direction of the radio axis after the telescope is set and adjusted will be different from its calculated direction.

1.2. Control of the Radio-Astronomic Receiving Apparatus

Simultaneously with the guidance of the radio telescope, the computer must provide for programmed control of the telescope's receiving apparatus. The two programs must be mutually consistent and coordinated: When observing, it is important to know what is being received at a given moment and from what direction.

The cosmic radio emission that is received is of a noisy nature. The quantities measured are the mean intensity of the noise emission over the entire band of radiometer reception (continuous-spectrum measurement) and the distribution of intensities over the spectrum within the band (spectral analysis). In the latter case, by means of frequency-spaced narrowband filters, the entire reception band is divided into frequency intervals in which intensity measurements are made. The measuring is done by a different method: The radiation that is received from the direction of the object being investigated is compared with radiation from a neighboring source.

A generalized diagram of the computer-controlled radio spectrometer, in which the intensity throughout the reception band is measured by one of the channels, is shown in Fig. 4. Variation of the direction of reception is achieved by the alternate connection of receiving antenna exciters A_1 and A_2 to the input of the amplification circuit. The differences of intensity between channels A_1 and A_2 are useful signals. These signals, after amplification, are released into the synchronous detector (SD) of each channel and, after a small analog-process buildup (on an RC ladder network), are connected through the commutator (C) to the electronic computer

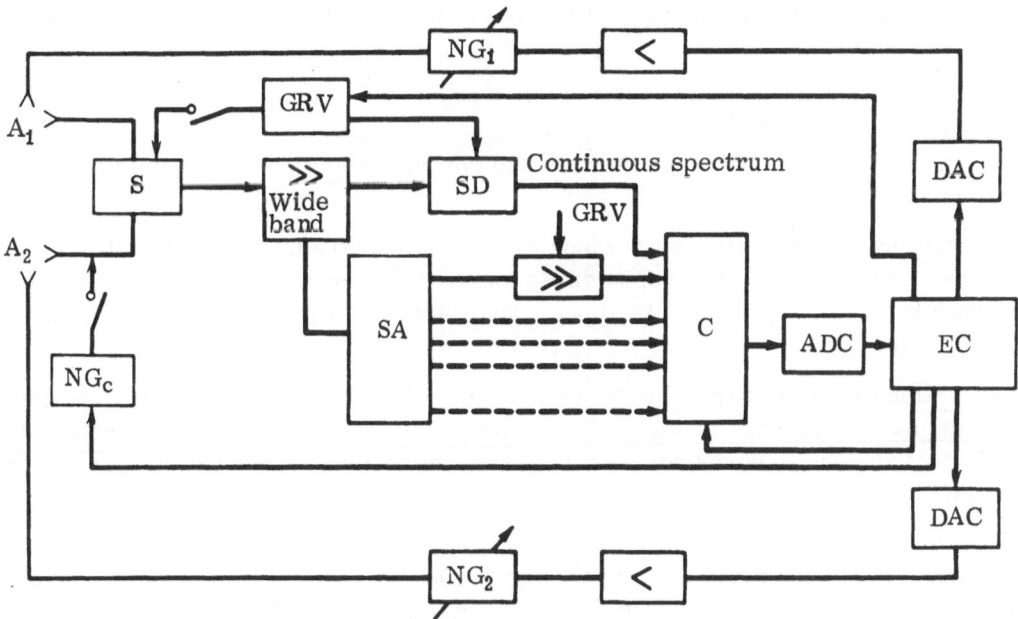

Fig. 4. Diagram of computer-controlled radio spectrometer. A_1, A_2, inputs of antenna exciters; S, switch; SA, spectrum analyzer; C, commutator; GRV, generator of reference voltages; SD, synchronized detector; ADC, analog-to-digital converter; DAC, digital-to-analog converter.

EC. The analog-to-digital converter (ADC) at the input of the computer converts the measurable quantities to the digital form.

With the measurements of mean intensity, the continuous-spectrum channel is connected to the computer. In spectral analyses, the narrow-band channels are connected in turn by the commutator to the computer. Since in all cases a very weak noise signal is received, the computer must provide for a sufficient accumulation for distinguishing the signal against the background of noise filtrations. Calibration of the radiation being received is effected by the connection of a calibration-noise-signal generator (NG_c) to the radiometer's input.

In order to ensure the stability of the apparatus' parameters over the wide band, it operates in the so-called null regime. With the help of auxiliary noise generators NG_1 and NG_2, whose emission is fed into channels A_1 and A_2, an equality of intensity is established at the latter's inputs. The accuracy of the balancing is determined by the agreement of the output level with the "null" level previously found with the disconnection of the generator of reference voltages (GRV).

Two balancing procedures are possible: irregular, done from time to time at the observer's discretion, and automatic, in which case the equality of the input intensities is maintained constant.

In this way, the electronic computer, incorporated in the radiometer circuit scheme, should ensure the execution of the following operations: (1) control of the commutator with successive connection of the narrow-band channels when making spectral measurements and with constant connection of the wide-band channel when making continuum measurements; (2) measurement of signals, their storage, and the determination of their mean value for some interval of time; (3) finding of the radiometer's "null" by switching off the GRV and, at the same time, the measurement of the output voltage of the wideband channel; (4) incidental balancing of the input intensities by control of the noise generators (NG_1 and NG_2) to obtain an intensity matching the previously found "null" level; (5) constant automatic balancing of input intensities at intervals of 1-5 sec; (6) calibration of the signals being measured by switching on the NG_c.

1.3. Observational Procedures

The following observational procedures and corresponding operating conditions of the apparatus are possible.

<u>Observations with a Stationary Antenna by the Drift Method.</u> The radio telescope is set on fixed A and Z coordinates. Measurement is made of continuum intensity as a function of time with an interrogation period of 1-10 sec. Sporadically, verification of "zero," balancing, and calibration are done.

<u>Observation by the Aiming—Withdrawal Method.</u> The radio telescope alternately tracks the source and some reference point. With this program it is possible both to measure the continuous spectrum and to carry out spectral analyses (the measurement of the continuum, single-channel reception of information, is a particular case of the multichannel spectral analyses).

The signal measured is the difference signal, which is determined in units of the calibration signal that is incorporated when the reference point is tracked.

The measurement-cycle system is shown in Fig. 5. The duration of a complete cycle of measurements is 8-20 min. Of this time 30% is tracking of the source and 50% is tracking of the reference point (including 20% for calibration). During the resetting of the antenna, no measurements are made. The measurement of the calibration level is initiated

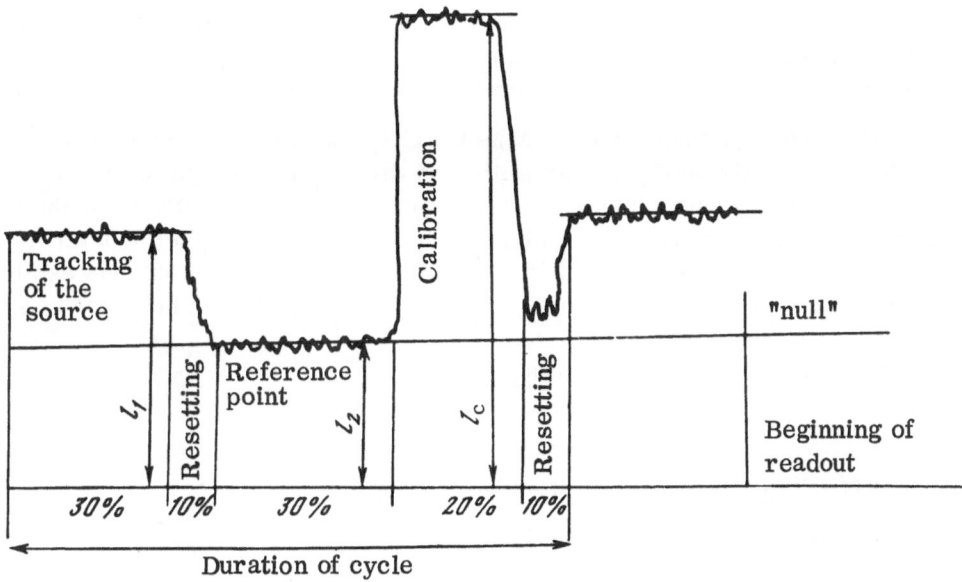

Fig. 5. Cycle of measurements by the aiming−withdrawal method.

30 sec after connecting the NG_c. Balancing is done either sporadically or automatically (operation by the null method). In the latter case, the continuum magnitude being measured is the value of the compensating noise signal.

Observation by the Aiming−Aiming Program. Observation by this program differs from the preceding method in that the angle between the object being investigated and the reference point is made equal to the angle between the two possible directions of reception. The experimental signal is alternately received by antenna exciters A_1 and A_2 with opposite deflections of the output levels from the established, standard "null." The algorithms used are those of balancing and observations by the aiming−withdrawal method; operation is by the null method. The strength of the signal, as compared to that observed by the aiming-withdrawal method, is double. Calibration is done in one of the two positions of the radio telescope, and the signal strength is determined by the equation

$$T_{ai} = \frac{l_{1i} - l_{2i}}{2(l_{ci} - l_{2i})} T_c.$$

Scanning. Intensity measurement is done in the continuous-spectrum channel with sampling every 2–10 sec. Balancing and temperature calibration are done before the beginning and at the end of the scanning cycle. The algorithms of observation by the drift method and the algorithm of balancing are used. Printed out are the measured value of the antenna temperature and the radio telescope's position at a given moment.

Adjustment or Positioning. This is done by using the radio emission of planets or other sources of small angular size, in the wide-band channel. When tracking a planet with the radio telescope, the system of aiming by means of the correcting unit includes corrections to azimuth, ΔA, and to zenith distance, ΔZ. In accordance with the program, first of all, by a given interval $b = 10'' - 60''$ the telescope is displaced in azimuth. After each interval or step, the mean signal value is found and compared with the preceding one. If the signal strength increases the telescope is shifted further in the same direction to the next step. In the contrary case a backward step is taken and the azimuth correction is concluded. If at the beginning of the correction a weakening of the signal is observed, then b changes sign and movement is made in the opposite direction. After the azimuth computation, an analogous program is carried out for Z. With the finding of the maximum signal strength in both coordinates, the telescope is deflected and the signal intensity is measured by the aiming—withdrawal method.

As a result of the adjustment, corrections to ΔA_{opt} and ΔZ_{opt} are found, with which the signal being received has its maximum strength, and this signal strength is found. The obtained values of the corrections are used for subsequent aiming of the RT-22 at the object being investigated, and the strength of the signal from the planet is used for determining the effective area of the antenna. Adjustment is carried out separately for antenna exciters A_1 and A_2 within the limits of correction values ΔA_{lim} and ΔZ_{lim}, taken to be equal to 4' × 4'.

2. The System's Configuration and Its Programmed Control

2.1. The Choice of a Type of Electronic Computer for the Complete Automation of Observations with the RT-22

The selection of a type of electronic computer for the automation of the observations involved the following requirements: (1) the computer must be sufficiently reliable for many hours of operation in the real time scale; (2) the computer must have extensively developed input/output channels and an interruption system; (3) the computer should be comparatively inexpensive so that its use may be expedient.

Taking into account that with the computer it is proposed to solve only the problems of aiming and observing, leaving the final processing to another electronic computer, one of the minicomputers that satisfy the above-enumerated requirements should be selected for installation on the radio telescope.

For the selection, the following computers were compared: the M-6000, the "Electronics-100," TPA/I (Hungary), and the "Videotone 1010B" (Hungary). The comparison was conducted in accordance with the following criteria: (1) accuracy of computation; (2) productivity (capacity and efficiency); (3) peripheral instrumentation; (4) provision for mathematical formulation.

Comparative Evaluation of Minicomputers for Their Accuracy of Computation and Productivity. The accuracy of computation and the productivity of the system are interrelated and therefore should be considered at the same time.

Both the problem of aiming the telescope and the storing of information require an accuracy surpassing that of all the enumerated computers. Therefore, for the presentation of the information it is necessary to use two or more computer words and organize the computations in subprograms. A number with multiple accuracy or extended precision may be represented in the form of a fixed, or floating, decimal point. Minicomputers usually have standard programs for operations with numbers represented in one of these forms or both. Under otherwise equal conditions, the representation of numbers with fixed decimal point is preferable, since the corresponding programs work faster.

For the problem of aiming, the required accuracy is determined by the accuracy of the measurement of coordinates and is equal to 21 binary orders. Table 1 presents data on the required number of computer words for each computer and the form of representation of the numbers for which there are complete sets of standard programs for this problem.

With storage, the number of binary orders outgoing for each channel is determined by the storage time. With a two-hour storage and with an interrogation cycle of 2 sec, not more

TABLE 1

Computer	No. of computer words	No. of binary orders	Form of representation of information
"Electronics-100," TPA/I	2	24	With fixed decimal point
	3	24	With floating decimal point (12-order, 24-mantissa)
Videotone 1010B	3	24	With fixed decimal point
M-6000	2	32	With fixed decimal point
	2	23	With floating decimal point (8-order, 23-mantissa)

than 23 binary orders are required. The necessary number of words to a channel, for the different computers, is found to be analogous to the problem of computing coordinates (see Table 1).

According to the criterion of the required accuracy, the M-6000 is preferable because it requires a smaller memory capacity for the representation of the information.

The criterion of computer productivity for the system being engineered is critical, since on the computer's productivity depends the aiming accuracy and the amount of information that is processed per unit time.

An evaluation of a computer's productivity according to the memory cycle does not give satisfactory results because it does not take into account the computer's architecture. The most accurate evaluation is provided by the method of expert programs. In this method, one or several problems most typical for the system are selected and sketchy programming is done for the computers being compared. The problem of information storage while observing a source is chosen as one such problem. A block diagram of a complete storage cycle for one channel is shown in Fig. 6.

As an expert program, the block chosen was number 2, which accomplishes the storage proper. Blocks 1, 4, and 5 characterize the system of input−output and of interruption, which can be evaluated by the tabular characteristics. At this stage a comparison was made of the minicomputers M-6000 and TPA/I. The "Videotone 1010B" computer was excluded from con-

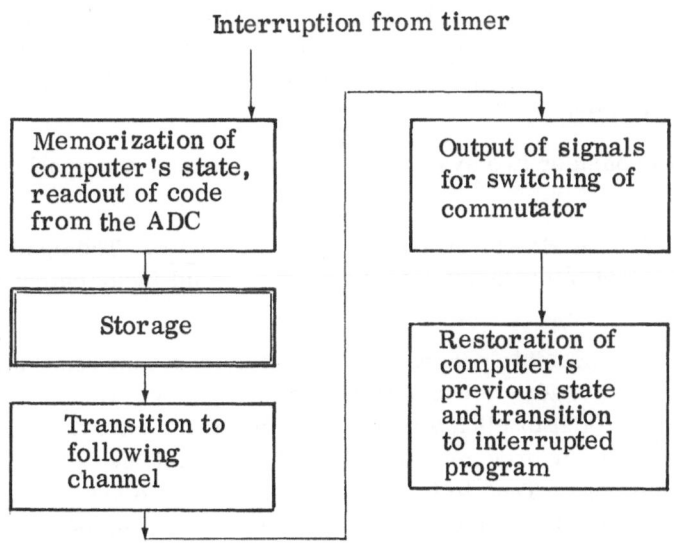

Fig. 6. Block diagram of storage cycle.

sideration because, under otherwise equal conditions, it requires a larger memory capacity and has reduced productivity. The "Electronics-100" was not considered because, with the equivalence of architecture, it has a longer memory cycle (2 μsec) than the TPA/I (1.5 μsec).

A comparison of these programs has shown that the M-6000's architecture is better than the TPA/I's because, in spite of the considerable difference in their memory cycles (2.5 and 1.5 μsec), the time for carrying out the program is approximately equal (72.5 and 67.5 μsec), and the number of memory elements required in the M-6000 is smaller than in the TPA/I (23 instead of 29). Furthermore, the M-6000 provides higher accuracy and has a more highly developed interruption system.

Comparison according to the Peripheral Equipment That Is Supplied and the System of Mathematical Support. A comparative analysis of the peripheral equipment that is supplied shows that the computers TPA/I and M-6000 have equipment of approximately equal value for satisfying the requirements of the system. The mathematical support, in spite of different organization, is also approximately the same.

On the basis of these evaluations, the M-6000 computer was selected for the automation of the RT-22 radio telescope [4, 5].

2.2. The System's Programming

The system's programming must ensure its operation in a real time scale with simultaneous solution of both the aiming problem and the information-gathering problem. Furthermore, there must be foreseen, and provided for, the possibility of operation by the experimenter with a system of mutual correlation of all these problems.

The programming must include the following components: (1) a real-time monitor for making the interruptions, controlling input/output, and determining the transmission or passing of the problems; (2) a set of programs for the aiming of the telescope that provide for the problems listed in Section 1; (3) programs for the control of the receiving apparatus and the collection of information; (4) programs for connection with the operating system and for the output of information in the specified format; (5) programs for the preliminary and the definitive processing of the information.

For the arrangement of all the possible programs in the computer's memory, it is necessary that there be a high capacity. This is inexpedient, since considerable capacity will be taken up by programs that are seldom used or not used at all in a specific experiment. Therefore all the backup programming should be stored outside for call into the programs's memory when the experimenter requires.

It makes sense to consider two systems of programming: the punched tape system and the disk system.

The Punched Tape System of Programming. The operation of this system requires a minimum amount of peripheral equipment: a rapid input/output device for the punched tape.

In this system all the possible programs are in a mobile form on the punched tape. The observer (the operator of the radio telescope) determines, in accordance with the established plan of experimentation, the list of all the programs needed for carrying out the experiment, including standard and auxiliary programs and the relations between programs, and he feeds the programs into the memory. Then to the monitor of real time are communicated the system's parameters: the initial data for the aiming, the starting time of the experiment, the frequency of the collection of information into the operator's instrument panel, etc. After this it is necessary to verify the correctness of the formulated set with a special test.

The punched-tape system has the following drawbacks: (1) the formulation of a complex program is time consuming; (2) mistakes are possible in the process of program formulation and their removal is difficult; (3) the system is inflexible: while carrying out the experiment it is practically impossible to make changes in the aiming algorithms or in the processing or the distribution of the results without reformulating the program; (4) a large memory capacity is required: it is necessary to store in the memory all the programs that may be required during the experiment, including those seldom used and programs not simultaneously employable.

The use of the disk system presents greater possibilities.

The Disk System of Programming. In this system a magnetic-disk memory (MDM) is used for program storage. The advantage of the MDM, its large capacity and short sampling or selection time, makes it possible to construct a highly effective and flexible system.

The formulation of a specific set of programs is done by command from the operator with a typewriter. The use of a magnetic disk as a systematic or methodical memory makes it possible: (1) to easily generate experimental programs at will; (2) to make changes quickly in the set of programs during an experiment; (3) to work with the system of each experimenter, since the relationship is conducted in operator-oriented language; (4) to considerably economize memory (computer storage) with a system of programs available to the experimenter: by command from the operator, any program can be called from the disk into the operative memory; (5) to operationally include, and defer, new programs in the system.

During the operation of the system, the programs are in an active or a passive state.

The programs considered active are those that are connected, by interrogation from the operator, to the comprehensive set of programs. These programs in turn are divided into those constantly in the memory and those normally on the disk.

Constantly in the memory are the programs often used (the time for access to them should be minimal): (1) specific programs of aiming and processing, selected by the operator in accordance with the plan of the experiment; (2) library programs used by the aiming and observation programs; (3) the resident part of the real-time monitor.

The programs normally located on the disk are called by the monitor into the memory, or computer storage, only when needed, if this does not create obstacles for the work of the basic programs. At the same time the appropriate library programs are also called forth. A special portion of the memory is assigned to these programs. The following programs are related to them: (1) the adjustment of the telescope and the balancing, adjustment, and calibration of the apparatus; (2) special display of information in the form of tables, graphs, and histograms with various peripheral devices; (3) several types of preliminary processing of the information; (4) the definitive processing of the information recorded earlier (if needed); (5) the translation and the resolution of problems into languages (if the procedure of joint observation and problem solving is being employed).

These programs have the least priority and are interrupted by all the other programs. By interrogation from the operator, any program can be shifted to the active state and also be called into the memory — into the computer storage. The structure of the system is shown in Fig. 7.

Required Capacity of Memory. Before the definitive elaboration and depositing of the programming, it is possible only to approximately evaluate the required capacity of memory.

The punched-tape system should include: (1) a basic control system (BCS); (2) input–output drivers; (3) a library of standard programs; (4) programs for carrying out the aiming

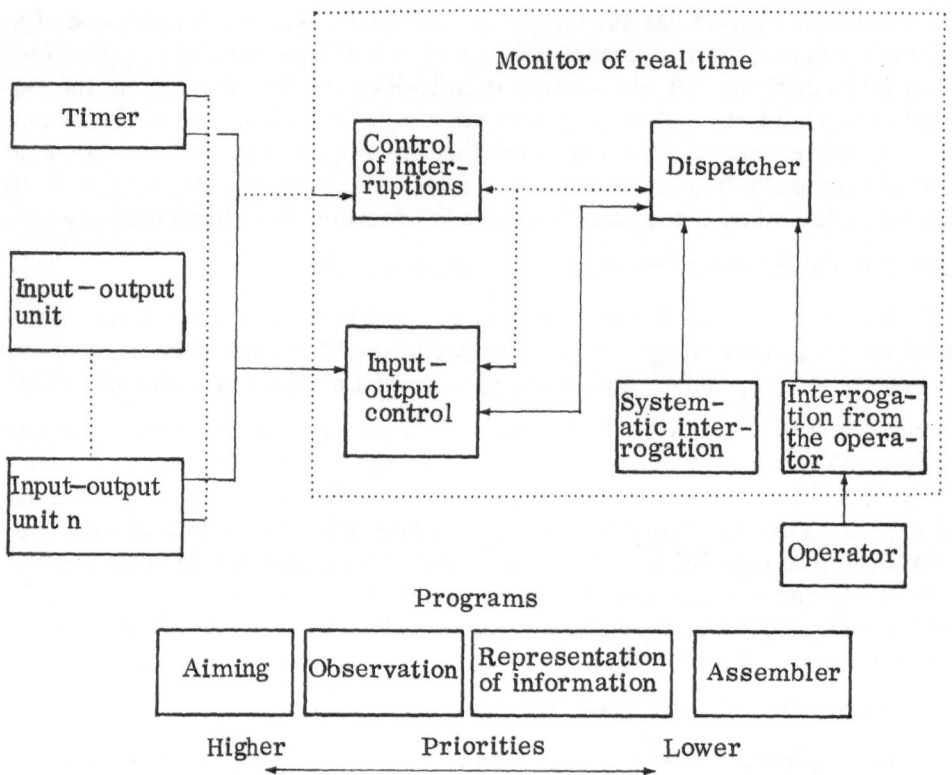

Fig. 7. The structure of the disk system.

procedures; (5) programs for storing experimental information, including memory fields and buffer zones for the input and processing of information; (6) programs for calibrating, balancing, and adjusting the radio telescope; (7) programs of communication with the operator of the system; (8) programs for organizing the output of data in the necessary format with a graph plotter, printer, oscillograph, or display.

The basic control system [6, 7] requires about 200 elements. The drivers of the keyboard printing unit (KPU), the punched-tape input unit (PTIU), the input—output unit (IOU), the punched-tape output unit (PTOU), the SID-1000, the graph plotter, and the oscillograph require 2500 elements [6, 7]. The drivers of the means of communication with the object and with the operator's control panel require also about 2500 elements. The programs for aiming by estimated calculations performed by algorithms require 4000 elements. The library of the required standard programs needs 600 memory elements [8].

The storage program requires 1000 memory elements; the memory field, 1000 elements; and the buffer field of the output, 300 elements. The programs for calibration, balancing, and adjustment by estimated calculations require 3000 elements. The programs of communication with the operator and with the organizations of output formats require not less than 1000 elements. All together, allowing for base pages and absolute charge or load, about 17,000 memory elements are required.

The memory capacity for the disk system of programming depends upon the existent standard disk systems of real time. The minumum capacity of memory for such a system is 16,000 elements.

Summing up the above, we conclude that the minimum capacity of memory necessary for automating the RT-22 is 16,000 elements (a minimal disk system and a punched-tape system with limited potentialities). However, provision should be made for a further enlargement of the memory up to 24,000 words.

2.3. Selection of the System's Equipment and the

Development of Its Configuration

The development of the system's configuration is based on the specific selection of its equipment with allowance for the specific properties or characteristics of the M-6000 layout.

The system's equipment includes the following units: (1) processor and computer memory; (2) external memory; (3) punched-tape equipment; (4) means of communication with the operator; (5) information output units; (6) information output and input units necessary for the guidance of the telescope; (7) input of information from the radio spectrometer; (8) output to the indicating instruments of the panel.

The processor must have productivity sufficient for solving the problems of aiming the radio telescope and for storing the experimental data. In view of the large volume of computations, the processor should be supplemented by an arithmetical expander (AE). As was determined in the preceding subsection, the minimum capacity of the system's memory is 16,000 elements. Therefore there must be four general memory units (GMU) of 4000 words each that can be connected to the processor through the memory increment unit (MIU).

External memory units are those based on magnetic disks (MDM) and on magnetic tapes (MTM). A magnetic-disk memory unit should, by its characteristics, ensure the efficient operation of the disk system. It is expedient to use a disk mechanism with a removable disk package, combined with a natural-synoptic electronic computer. A magnetic-tape memory unit is applicable in the system for many-hour storage of experimental information for the purpose of subsequent storage and processing in larger and more powerful computers. Therefore the memory unit must be compatible with the memory units of the powerful computers. In view of the wide incorporation of the natural-synoptic electronic computer, the MTM must be natural-synoptic compatible.

For the efficient operation of the punched-tape system, the assembly should include a fast input—output information mechanism (a photoelectric reader and a tape perforator).

For operative input and output of information, there must be a mechanism like a typewriter. In the complete set of accessories there is provision for a keyboard printing unit (KPU) based on a "Konsul-260" typewriter. An additional device that may be used is the input—output unit.

For active interaction between the operator and the system for representing operative information and observational results, the system utilizes an SID-1000 data-indication station. The equipment also includes an oscillographic display unit (ODU) that provides graphic display on the screen of a standard oscillograph, and a TsGP-1 digital-to-graph unit for graphical display of the observational results.

For the radio telescope's guidance, it is necessary to provide for the input of information on the telescope's position from angle-to-digital data units (2×18 bits) and the output of control actions in analog form. For the input of the information from the angle-to-digital data units, the equipment must include a discrete information input module (DIIM) with appropriate framing. The output of the analog control actions is accomplished by the digital-to-analog converter (DAC).

The input to the electronic computer from the radio telescope is effected through an analog-signal input subsystem that feeds the computer the continuous spectrum of the radio-meter and up to 255 narrow-band channels from the spectrum analyzer.

The output of the channels is connected to the commutators. In the aggregate M-6000 system, there are two types of signal commutators: the mid-level signal commutators (MLSC) and the low-level signal commutators (LLSC). All together, there are 16 commutators in the subsystem (eight for the low level and eight for the high level), and each commutator is designed for connection to 16 channels. The commutators are controlled by commutator control modules (CCM) which are parts of the input—output expander (IOE). From a mid-level commutator, an analog signal is sent through a CCM to the fast-acting analog-to-digital converter (ADC), going out also to the IOE. From a low-level commutator, an analog signal (also through a CCM) is sent to the low-level signal amplifier (LLSA) and to the analog-to-digital converter with integration [ADC(i)].

In view of the large amount of information that is received with spectral radio-astronomic observations and the possible difference of radiospectrometer design, it is advisable to use commutators of both types.

A system configuration developed on the basis of the above-mentioned considerations is presented in Fig. 8. The configuration was developed on the basis of data on the logical layout of the M-6000 system [9]. At the present time the supplying of the M-6000 is being done on the basis of type assemblies [10]. For the whole system, assembly No. 4 was selected. Modules not included in this assembly are shown in Fig. 8 by dashed connecting lines.

While the RT-22 automation project was being carried out, tests were conducted on two systems of partial automation using the electronic computers TPA/I [11] and M-6000 (in a

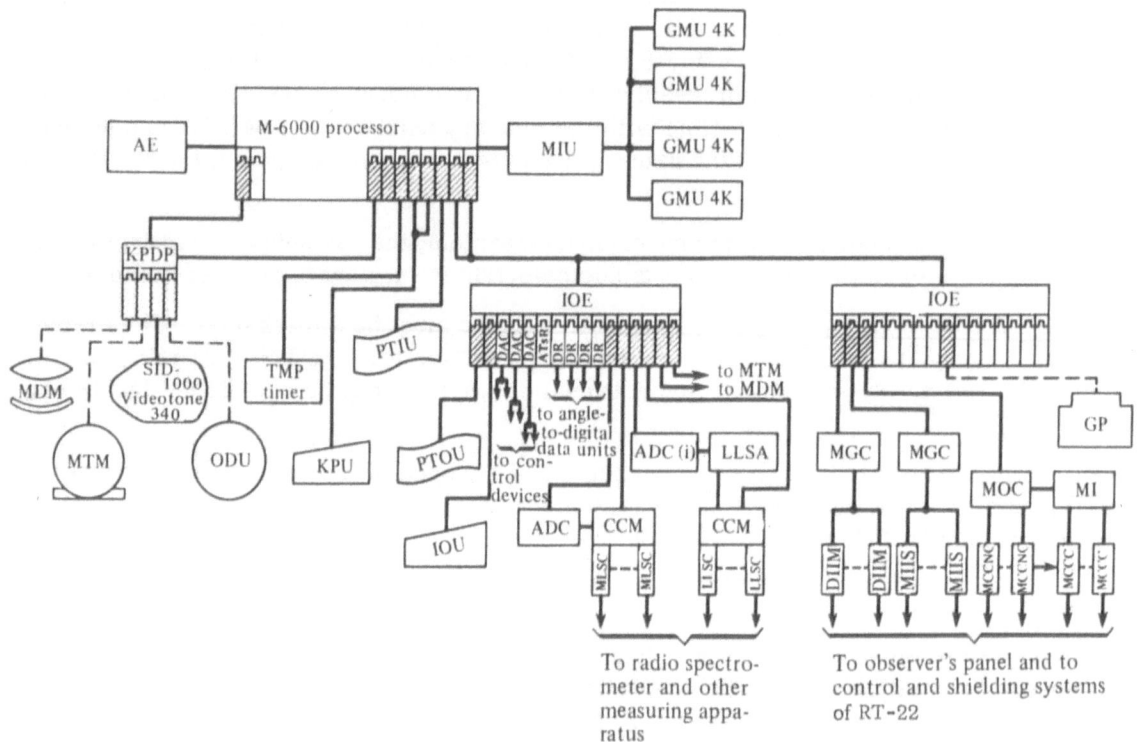

Fig. 8. Configuration of the M-6000 system for the RT-22. MIIS, modules for input of initiative signals; MCCNC, modules for code control, noncontact; MCCC, modules for code control, contact; MGC, module for group control; MOC, module for output control of discrete information; MI, module for increment; DR, duplex register. For other abbreviations, see text.

minimal configuration) [12], which made it possible to determine more accurately the requirements and to develop a series of programs. The experiments that were conducted confirmed the correctness of the solutions that had been selected and their urgency at this time.

The authors thank G. N. Kuklin for his discussion of this report and I. I. Berulis for his help in the formulation of algorithms.

LITERATURE CITED

1. P. D. Kalachev and A. E. Salomonovich, Radiotekh. Élektron., No. 3, p. 4 (1961).
2. V. A. Vvedenskii, P. D. Kalachev, A. D. Kuz'min, Yu. N. Semenov, and R. L. Sorochenko, Tr. FIAN, 93:45 (1977) (present volume).
3. V. M. Gudnov and R. L. Sorochenko, Astron. Zh., 49:1001 (1967).
4. V. V. Ryazanov, V. G. Vinokurov, I. I. Itenberg, V. M. Kostelyanskii, and G. M. Lekhanova, Tr. Nauchn. Issled. Inst. UVM, No. 2, Severodonetsk (1970).
5. Architecture of the M-6000 [in Russian], Severodonetsk (1972).
6. SPO−6000A. General Description [in Russian], Severodonetsk (1972).
7. SPO-6000 A. Basic Control System [in Russian], Severodonetsk (1972).
8. SPO−6000A. Library of Standard Programs [in Russian], Severodonetsk (1972).
9. Logical Layout of System Based on Processor M-6000 ASVT-M [in Russian], Severdonetsk (1972).
10. Nomenclature List of NPO "Impulse" Publications [in Russian], Severodonetsk (1973).
11. A. V. Kutsenko, B. A. Polos'yants, G. T. Smirnov, R. L. Sorochenko, and S. A. Terekhin, Preprint FIAN, No. 84 (1974).
12. T. I. Alfimova, A. V. Kutsenko, B. A. Polos'yants, G. T. Smirnov, R. L. Sorochenko, S. A. Terekhin, and V. A. Shirochenkov, Izv. Vyssh. Uchebn. Zaved., Radiofizika, 19(10): 1469 (1976).

THE PROBLEM OF MATCHING A NOISE GENERATOR

I. A. Alekseev and V. N. Brezgunov

The problem of matching a noise generator based on a 2D2S diode in the meter-wavelength band with the help of a low-frequency filter section is examined. The circuit of a powerful noise generator with a ferrite impedance changer is described. The possibility of matching this circuit in a wide band of frequencies is shown.

As shown in [1], the formulas for determining the noise coefficient of an apparatus being tested and the temperature developed by a noise generator under load are derived under the assumption of an ideal matching of the noise generator. Mismatch leads to the occurrence of two undesirable effects.

Firstly, the power delivered by the generator to the load is not known exactly and, consequently, an error is introduced into the quantity being measured by means of the noise generator. In [2] it is shown that for achieving an accuracy of measurement of ±2% it is necessary that the traveling-wave ratio (TWR) of the noise generator be 0.98.

Secondly, in the case of a rather long feeder tract, or channel, included between the noise generator and the quadripole being tested (which has some input reflection coefficient), in that channel there occurs a complex interference pattern [3]. In a general form, the interference terms that are determined by the natural noises and those of the noise generator, in the input section of the quadripole being tested, are

$$\Delta T_{i1} = T_1 \rho_2^2 \eta^2 \frac{1}{M} \cos 2\beta l,$$

$$\Delta T_{i2} = T_2 \rho_1^2 \eta^3 \frac{1}{M} \cos 2\beta l,$$

where T_1 and T_2 are the noise temperatures of the generator and of the input circuits of the quadripole being tested, respectively; ρ_1 and ρ_2 are the reflection coefficients at the input of the quadripole and at the output of the noise generator; η is the efficiency of the feeder; M is a certain function that depends on the transmission band of the quadripole and the length of the feeder, and is called the "visibility function"; $\beta = 2\pi/\lambda_F$ is the imaginary part of the propagation constant (λ_F is the wavelength in the feeder).

The existence of a frequency-dependent interference pattern leads to errors in the determination of noise-coefficient and noise-temperature values, since the levels of the natural noises and of those from the noise generator at the input of the quadripole being tested are determined by the length of the feeder tract, or channel, and by the frequency adjustment and the transmission band of this quadripole.

The latter effect acquires particular importance in the solution of problems such as the determination of the efficiency of long feeder systems and the determination of the noise coefficient in ideally matching systems having a long line of transmission to the input, in the conducting of absolute measurements in radio astronomy, etc.

In the present paper we examine the problem of matching a generator of noise in the meter-wavelength band, in which a 2D2S diode is used as the primary noise source. This problem is discussed in [1, 4].

Let us point out briefly that the idea of matching proposed in these publications consists in the neutralization of the diode's output capacitance and that of the assembly by including them in a resonance circuit. With this method the bandwidth in which matching is achieved is limited by the capacitance of the assembly and the resistance of the load.

In [1] it is shown that with the tuning of the circuit to resonance at a frequency of 50 MHz, a noise generator works satisfactorily in a range of 20-30 MHz with $R_L = 75$ Ω. The tuning of a low-quality or low-Q circuit when measuring in a wide-frequency band is difficult.

In the fabrication of a noise generator for operation in the meter-wavelength band (to 150 MHz), its design is often carried out as a noncoaxial variant [1, 4]. In such a design, the matching can be done by means of a low-frequency filter section with the inclusion of the parasitic reactances in this filter. With this, there is no longer any need of retuning during operation in a wide band of frequencies, and the construction is substantially simplified. By a suitable selection of the filter section's elements, it is possible to ensure good matching up to frequencies of the order of 150 MHz. Let us consider the noise generator substitution circuit (see Fig. 1).

The inductance of the diode's anode leadout (L_D) and the diode's anode−cathode capacitance C_{ac} (or C_D) constitute a series oscillating circuit having its resonance at a frequency of the order of 3 GHz [1]. Consequently, at the frequencies of the band under consideration, the circuit's resistance is of a capacitance nature. If we take into consideration the assembly's capacitance, then the total capacitance, parallel to the resistance of the anode load (R_L), comprises a quantity of the order of 5 pF.

With the specified execution of the structural plan, the conductor that connects the output joint to the diode has its own inductance of the order of 0.03 μH. The inclusion of a supple-

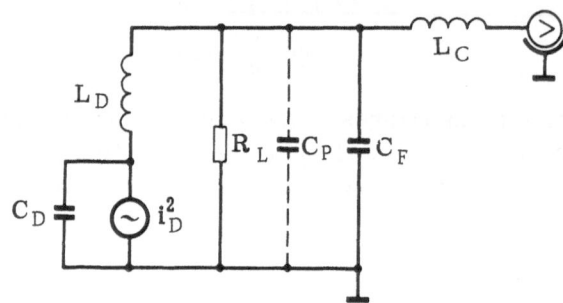

Fig. 1. Noise generator substitution circuit. L_D, diode's anode−leadout inductance; C_D, diode's anode−cathode capacitance; R_L, anode−load resistance; C_p, distributed parasitic capacitance of assembly; C_F, capacitance of filter; L_C, inductance of mounting conductor that connects diode's anode leadout to generator's output joint; i_D^2, square of the effective value of the diode's noise flux.

mentary capacitance C_F = 5 pF gives an inverted-L-shaped (Gamma-shaped) "K"-type filter section with a wave impedance

$$\rho = \sqrt{L_C/(C_D + C_P + C_F)} \approx 75 \ \Omega.$$

The critical frequency of such a filter

$$f_{cr} = \frac{1}{\pi \sqrt{L_C(C_D + C_P + C_F)}} = 400 \ \text{MHz}$$

is considerable higher than the operating band (150 MHz), which ensures a good constancy of filter wave impedance in the band and, consequently, also ensures good matching.

It has been experimentally determined that the TWR of a noise generator matched by this method and having its R_L = 75 Ω in a range of 20-150 MHz is not under 0.97 [5].

In many cases it is necessary to have a powerful initial source of noise temperature. With the use of a diode noise generator for this purpose, the designers increase the anode-load resistance.

The interelectrode capacitance and the assembly's parasitic capacitance can be incorporated in a low-frequency filter also in the designing of noise generators having powerful outputs.

For the changing of impedance in a powerful noise generator, exponential, step, and other types of changers can be used. However, for the meter-wavelength band it is most expedient to use an automatic-changer circuit based on ferrite [6]. With this, the inductance of the changer's dissipation and the output capacitance of the diode, together with the assembly's capacitance, constitute a low-frequency filter section. The basic circuit of the diode noise head's anode loop is shown in Fig. 2.

Experimental testing of a noise generator with R_L = 300 Ω has shown that the circuit is matched with a wave impedance of 75 Ω with a TWR not under 0.97 in a range of 20-150 MHz. The autotransformer's (or impedance changer's) efficiency is of the order of 0.95-0.98 and only slightly depends on the frequency. The noise temperature at the output of such a noise generator is determined by the formula

$$T = 435N^2 I_{ng} \eta(f),$$

where T is in degrees Kelvin, N is the coefficient of transformation, I_{ng} is the current of the noise generator, mA; $\eta(f)$ is the efficiency of the autotransformer (or impedance changer) as a function of frequency.

The use of the proposed method of matching in a wide range of frequencies and the use of ferrite impedance changers (or autotransformers) makes it possible to obtain matching with a wave impedance of 75 Ω and with higher values of the noise diode's anode load.

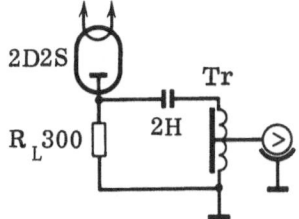

Fig. 2. Basic circuit of noise generator's anode loop. Tr is an autotransformer with a core 4.5 × 1.5 × 4.5 of 2000 H.

LITERATURE CITED

1.   A. D. Kuz'min, Measurement of the Noise Coefficient of Receiving-Amplifying Apparatus [in Russian], Gosénergoizdat (1955).
2.   J. Harris, Proc. Inst. Electr. Eng., 108:103 (1961).
3.   V. S. Troitskii, Zh. Tekh. Fiz. 25:8, 1426 (1955).
4.   N. M. Teterich, Noise Generators and the Measurement of Noise Characteristics [in Russian], Énergiya, Moscow (1968).
5.   V. N. Brezgunov and V. A. Udal'tsov, Prib. Tekh. Éksp., No. 6, p. 100 (1968).
6.   C. L. Ruttroff, PIRE, 47(8):1337 (1959).

SOME STRUCTURAL DESIGNS OF THE MAIN SUBASSEMBLIES OF FULLY INCLINABLE AND ROTATABLE PARABOLIC RADIO-TELESCOPE ANTENNAS

P. D. Kalachev, I. A. Emel'yanov, V. P. Nazarov, V. L. Shubeko, and V. B. Khavaev

Optimal structural designs of mirror systems of fully inclinable and rotatable radio telescopes intended for operation in the meter-wavelength band are examined. The work of constructing a parabolic mirror with a reflecting mesh surface and with a flexible parabolic framework made up of cable and wire elements is analyzed, as also the principal subassemblies of a radio telescope: inclinable and rotatable supporting structures based on a floating pontoon or on an air cushion. Elevation-angle inclining apparatus with a traction carriage is also examined.

INTRODUCTION

For the solution of radio-astronomical problems that are becoming complicated, it is necessary to design and build very large radio telescopes. The increase in the size of radio telescopes is related to the rapid rise in their cost, in particular of radio telescopes with fully inclinable and rotatable parabolic antennas, that is, parabolic mirrors. Thanks to their universality, such radio telescopes have come into the most widespread use; therefore we shall examine precisely such radio telescopes.

The size of a radio telescope is determined by the size of its parabolic mirror, and any increase in the size of the latter, in turn, is related to an increase in the size of the other principal subassemblies: the structure of the mirror's suspension, supporting structure, inclinable and rotatable supporting structure, and drive mechanisms. There is also an increase in the power of the electric drive motors.

The high cost of radio telescopes in general and of radio telescopes having fully inclinable and rotatable parabolic antennas in particular is a serious obstacle to the increase in their size.

The cost of building a radio telescope depends, of course, not only on the size of the parabolic antenna. It depends just as much, or more, on the accuracy of fabrication required for the reflecting surface of the parabolic mirror, and this depends on the wavelength to be used. The shorter the wavelength with which the radio telescope is to operate, the more accurate must the mirror's fabrication be. The fact is that the effective area of the mirror's aperture, or its utilization coefficient, falls off rapidly with increase in the deviation of the mirror's actual surface from parabolic.

Decrease in the effectiveness of the mirror's area with decrease in the accuracy of fabrication of its reflecting surface is described by the well-known equation [1]

$$\eta_S = \exp\left[-(4\pi\sigma/\lambda)^2\right],$$

23

where λ is the wavelength and σ is the root-mean-square error of the mirror's reflecting surface.

This root-mean-square error σ is the total error; it includes error in the mirror's fabrication (technological error) and errors due to the mirror's elastic deformation (from its own weight, the wind, and temperature). The technological errors of fabrication and the errors resulting from deformations also depend on the size of the mirror.

Thus, a radio telescope's construction cost increases with increase in the size of its antenna and also with decrease in the wavelength being used. It must be noted that the rise in radio-telescope cost with increase in the size of the parabolic mirror, when using the short waves (for example, the centimeter and especially the millimeter bands), takes place considerably faster than when using the long waves (the decimeter and especially the meter bands).

This is explained by the fact that the cost of the antenna of a radio telescope used in the meter band is virtually independent of the parabolic mirror's accuracy of fabrication, the mirror in this case being fabricated with the tolerances usual in the building of metal structures.

With fully inclinable and rotatable radio telescopes for the meter band, it is possible to solve certain important problems of radio astronomy such as the observation of discrete sources, the Galactic halo, pulsars, etc.

Below will be examined some structural designs of the fully inclinable and rotatable parabolic antennas of radio telescopes intended for operation in the meter-wavelength band.

1. THE PARABOLIC ANTENNA (MIRROR)

A parabolic antenna is usually built according to a radially symmetric design. This means that the principal load-bearing elements of the framework of such a mirror are positioned in radial directions with respect to the center of the mirror. These radial elements of the framework are interconnected by chord elements that form the multiangular rings of the frame. Each one of these rings connects the radial elements at points equidistant from the center and is all in one plane perpendicular to the mirror's geometric axis.

A parabolic mirror's frame, since it contains curved elements that are near-parabolic, is a rather complex three-dimensional structure and, consequently, difficult to build and costly.

In the case of the parabolic antennas of radio telescopes intended for the meter band of wavelengths, the reflecting surface is usually made of wire grid (mesh or screen) or of perforated metal sheet (for example, sheet aluminum) [2-4]. The weight saving achieved by such methods of construction is accompanied by a considerable reduction of the radio telescope's total cost. However, the load-bearing framework that gives the parabolic shape to the reflecting surface of the main mirror has, up to now, continued to be comparatively heavy and complicated both in manufacturing and in construction.

Vitkevich and Kalachev [4] have described the structural design of the fully inclinable and rotatable parabolic mirror of a large radio telescope, the reflecting surface of which is formed of wire mesh; the framework for giving the parabolic shape to this reflecting surface is made up of a system of mutually parallel and perpendicular elements of cable and wire and a system of bracing wires.

While having several advantages, the design described in [4] has one vital defect: the antenna structure does not ensure matched deformations [5, 6]. In the present case, by "matched deformations" we mean that with certain stiffnesses of the load-bearing frame, and tightnesses of the wire elements that suspend the mesh of the reflecting surface from the load-bearing frame, their combined deformations produced by their own weight are such that the original parabolic shape of the main mirror is maintained invariant with any orientation of the mirror system in space. The design described in [6] is one with a radially equilibrated mirror. This main mirror, suspended from the load-bearing frame by radial elements, consists of a stiff metallic frame with a continuous sheet facing. This radio telescope is intended for operation in the short-centimeter-wavelength band.

In the proposed new structural design of a fully inclinable and rotatable parabolic antenna intended for operation in the meter-wavelength band, advantage is taken of the designs described in [4] and [6].

A new design (Fig. 1) is a structure with a spatially equilibrated wire-mesh mirror, the reflecting surface of this main mirror being formed by a wire mesh. The parabolic shape of the wire mesh is determined by a parabolic frame formed by a system of radial and chord cable elements and also by a system of radial bracing wires and a system of bracing wires parallel to the mirror's geometric axis. The radial cable elements that form the parabolic frame are fastened at their outer ends to angle brackets of the load-bearing frame, located on its peripheral ring. The load-bearing frame, all the way to the angle brackets, is a simple plane-parallel three-dimensional girder consisting of flat radial and chord trusses of constant structural height. The radial cable elements of the parabolic frame coincide (in the top view) with the directions of the load-bearing frame's radial elements, so that each radial cable element is located in the plane of the corresponding flat truss of the load-bearing frame.

The radial cable element of the parabolic frame is divided into equal (in the projection) segments, through the ends of which pass chord cable elements. At the points of intersection of the radial and chord cable elements are fastened the bracing wires that comprise the system of radial bracing wires and the system of bracing wires parallel to the mirror's geometric axis.

With the horizontal position of the plane of the mirror's aperture, the radial cable elements, being extended in conformity with the action of their own weight and that of the struc-

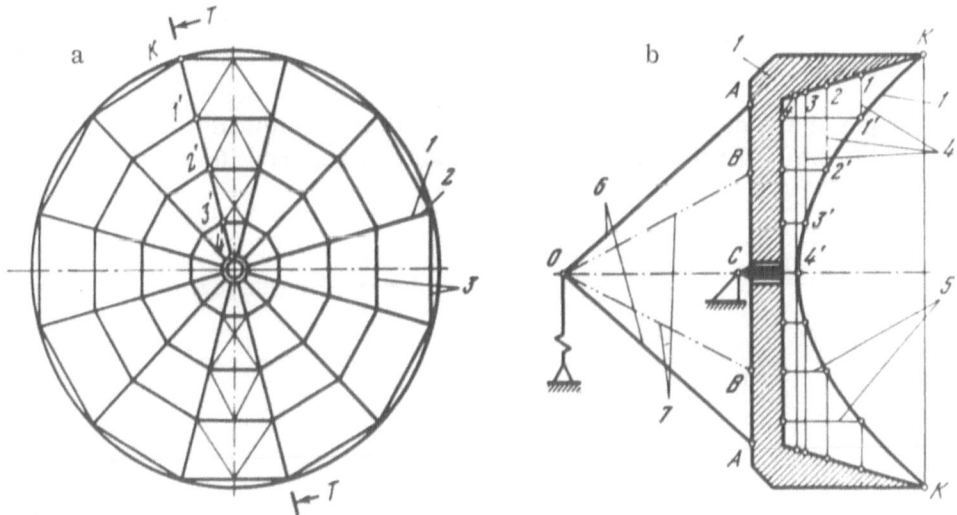

Fig. 1. Design of the parabolic antenna (principal mirror) of a radio telescope with a parabolic framework of cables. a) Top view; b) cross section through radial element T−T; 1) radial element (flat radial truss) of the mirror's load-bearing frame; 2) radial cable element of the parabolic frame; 3) chord cable elements; 4) bracing wires of the radial system; 5) bracing wires parallel to the mirror's geometric axis; 6, 7) supporting bar pyramid of the main mirror's suspension bracket.

tural elements attached to them, adopt the shape of a curve that closely approximates a parabola (a catenary). With the antenna turned toward the horizon (with the vertical position of the plane of the mirror's aperture), a radial cable element situated in or near a vertical plane, if it were fastened only at its ends, would lose its original parabolic curve under the action of its own weight. This, of course, is a problem that concerns all the radial cable elements of the parabolic frame. We specified a vertical element only because this problem is more obvious for it (see Fig. 1). For a cable fastened in space at a certain number of points uniformly distributed along it, the above-mentioned loss of the original parabolic shape can be obviated. However, in order to completely maintain the original shape of the cable element with any spatial orientation it may have, it would be necessary to secure it at an infinite number of points uniformly distributed along its entire length. In practice, the number of points of spatial securing of a cable is finite and is determined by the prescribed accuracy of the mirror's reflecting parabolic surface. The spatial securing of a radial cable element, at each point, is effected in three mutually perpendicular directions: radial, parallel to the mirror's geometrical axis, and tangential. In the first two directions the securing is done by bracing wires 4 and 5 respectively, and in the tangential direction by chord cable elements 3 (see Fig. 1).

If we assume that the securing point of a radial cable is immobile, then the distance between neighboring points of securing is determined by the equations

$$l = 2\sqrt{2fN/q}, \tag{1}$$

$$N = S_c \, [\sigma], \tag{2}$$

$$f + \delta = \lambda/12, \tag{3}$$

$$\delta \approx (l_{1-k})^2/16F_f, \tag{4}$$

$$l_{1-k} = l/\cos \varphi, \tag{5}$$

$$q_\Sigma = \gamma S_c \, k, \tag{6}$$

where f is the depth of arc of the deflection in the radial cable element in a length l which is the projection of the element between neighboring points of securing (in centimeters); N is

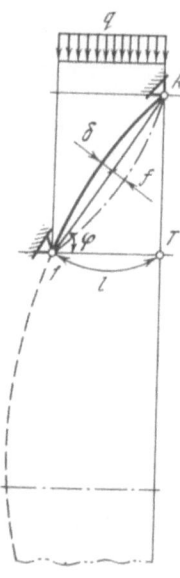

Fig. 2. Diagram of the securing of a radial cable which forms a parabolic cable frame of the mirror (the uppermost section when the antenna is directed toward the horizon).

the tensile stress in the radial cable (in kilograms); λ is the shortest wavelength at which the projected radio telescope can operate with the specified efficiency, that is, with the specified coefficient of utilization of the mirror's aperture area (CUA); S_c is the cross-sectional area of the radial cable; γ is the specific weight of the radial cable's material; $k = q_\Sigma/q_c$ is the coefficient of loading or load factor, on the radial cable, where q_Σ is the total linear weight loading on the cable (in kg/cm) and q_c is the linear weight of the cable; l_{1-k} is a section (in centimeters) of the radial cable between points of securing (Fig. 2); F_f is the focal distance of the parabolic mirror; and δ is the depth of arc in the deflection of the segment of parabola (in centimeters).

For an evaluation of the length l let us consider, for example, a radio telescope with a 100-meter parabolic (and fully rotatable) mirror intended for operation in the meter-wavelength band. Let the shortest wavelength at which this radio telescope will operate with an acceptable CUA be $\lambda = 100$ cm. With a safety factor $n = 9$ and a temporal or time resistance of the cable's material $\sigma_t = 18,000$ kg/cm^2, the safe tension will be

$$[\sigma] \approx \sigma_t/n = 2000 \text{ kg/cm}^2. \tag{7}$$

Preliminary calculations show that the load factor $k \simeq 10$; and the focal distance of a radio telescope's parabolic mirror is usually within a range of $F_f = (0.3-0.4)D_m$, where D_m is the diameter of the mirror. For a steel cable $\gamma = 7.85 \cdot 10^3$ kg/cm^3. From Eqs. (1) to (6) we have

$$l = \cos\varphi \cdot \sqrt{\frac{\lambda F_f [\sigma]}{[\sigma] + 2\gamma k F_f \cos^2\varphi}}. \tag{8}$$

Assuming $F_f = 0.35D_m = 35$ m, $\varphi = 60°$, and substituting the appropriate values in Eq. (8), we find for our specific case $l \approx 277$ cm and $l_{1-k} \approx 554$ cm.

The total deviation of the mirror's reflecting surface from the specified parabola is determined by Eq. (3)

$$(\delta + f) = \lambda/12 \approx 8.32 \text{ cm.}$$

For a value of $l = 277$ cm, Eqs. (4), (5) give us $\delta \approx 5.48$ cm and $f = 8.32 - 5.48 \approx 2.84$ cm. From Eq. (1) and considering $f = 2.84$ cm and $l = 277$ cm, we find

$$N = 33.7q, \tag{9}$$

but $q = q_c k/\cos\varphi \approx 20q_c$ or $q_c \simeq 0.05q$. Here the total linear load on the cable, q, depends mainly on the wavelength λ and to an insignificant degree on the diameter of the main mirror.

Let us examine two cases of cable loading q: (a) $q_{uniced} \simeq 6$ kg/m (without ice deposit), and (b) $q_{iced} \simeq 50$ kg/m (with ice deposit).

For the first case, $q_1 = q_{uniced}/\cos\varphi = 12$ kg/m. From Eq. (9), $N_1 = 404$ kg. By using Eq. (7), we find the cross-sectional area of the cable from Eq. (2), $S_c = 0.202$ cm^2, and the diameter of the cable allowing for the organic core is

$$d_c \approx 2c\sqrt{S_c/\pi}, \tag{10}$$

where $c = 1.45$ is the coefficient that takes into account the effect of the core; $d_c \approx 7.5$ mm.

For the second case of cable loading, $q_2 = q_{iced}/\cos\varphi = 100$ kg/m. In order to determine N_2, we will use a well-known relationship [7]:

$$N_2^3 + \left(\frac{ES_c q^2 l^2}{24N_1^2} - N_1\right)N_2^2 - \frac{ES_c q_2^2 l^2}{24} = 0, \tag{11}$$

from which we find $N_2 \approx 1160$ kg. Then we find the tension in the material of the cable by using Eq. (2): $\sigma_2 = 5470$ kg/cm^2. Thus, the safety factor is n $= \sigma_t / \sigma_2 \approx 2.8$.

When there is complete icing, observations are not conducted and measures are taken to remove the ice.

The selected section of radial cable must be checked by computation of the cable's strength as a flexible filament over the entire span, equal to the mirror's diameter, and of the same two cases of loading q_1 and q_2 with allowance for the loads from the vertical bracing wires when the mirror is directed toward the zenith, that is, of the loads: $q_1' = q_1 + \Delta q$ and $q_2' = q_2 + \Delta q$, where $\Delta q = 2$ kg/m, the additional loading from the vertical bracing wires.

In our specific case, $D_m = 100$ m, and the mirror's h $= f = 17.85$ m, and

$$N_1' = q_1' l^2/8f \approx 560 \text{ kg.}$$

Using Eq. (1), we find $N_2 \approx 3500$ kg. The tension in the material of the cable will be $\sigma_2 = 3500/0.202 \approx 17,100$ kg/cm^2, and the safety factor n $= \sigma_t/\sigma_2 \approx 1.05$.

Complete icing of the mirror's mesh is an improbable case; therefore the above-derived safety factor is sufficient.

Thus the parabolic frame that gives the parabolic shape to the reflecting mirror surface formed of wire mesh can be a system of radial and chord steel cables, three-dimensionally secured at points spaced at equal distances in their projections on a plane parallel to the plane of the mirror's aperture. For example, for a radio telescope with a parabolic mirror 100 m in diameter and focal distance of 35 m, intended for operation at a minimum wavelength $\lambda_{min} = 100$ cm, the intervals between cable-securing points are 554 cm (in the above-mentioned projections), the diameter of the steel cable being about 8 mm. We have derived these values under the assumption that the cable-securing points are immobile. However, the mobility of these points is such that the parabolic shape formed by them is maintained with any orientation or inclination of the mirror, that is, there is no shifting of the junction points with respect to one another, which is equivalent to their immobility. This is achieved by matched deformations of the mirror, on which principle the design under consideration has been developed.

2. INCLINABLE AND ROTATABLE SUPPORTING STRUCTURE

The inclinable and rotatable supporting structure of a radio telescope's mirror is intended to absorb weight loads and to make it possible to aim the mirror-antenna at any point of the visible celestial hemisphere.*

The inclinable and rotatable supporting structures of existing radio telescopes differ, first of all, in the arrangement of the antenna's rotational axes. The principal designs are those with azimuthal-elevation-angle and equatorial arrangements of the antenna's rotational axes. Designs with other arrangements of the rotational axes are seldom used.

In the large radio telescopes where there is a sharp increase in the constructional and technological difficulties for putting into practice the design with equatorial arrangement of the rotational axes, the design with azimuthal-elevation-angle arrangement of these axes is obviously the only expedient one.

The structural designs with the azimuthal-elevation-angle arrangement of the rotational axes are, in turn, of two types: with a vertical base and with a horizontal base. In the first

---

*We recall that we are talking about fully inclinable and rotatable radio telescopes.

case the moment from transverse loads (wind loads) is absorbed by two supporting bearings situated at different levels on a vertical axis of rotation (a vertical column-turret). In the second case the moment is absorbed by a horizontal platform with supporting points in the form of balls, rollers, or individual dollies or carriages distributed uniformly around a rolling circle.

For very large radio telescopes, for example those having a parabolic mirror of the order of 100 m in diameter or more, it is evidently impractical to use an inclinable and rotatable supporting structure with a vertical base. In our opinion, for such radio telescopes the only arrangement, at least at this stage of development, is the structural type that has a horizontal base.

At the present time there are radio telescopes that have main mirrors 90 and 100 m in diameter [8, 9], and even larger ones are planned. The inclinable and rotatable supporting structure of such radio telescopes has the azimuthal-elevation-angle arrangement of the rotational axes, with a horizontal base in the form of a turntable on carriages [10].

The drawback of the rotatable support on a horizontal base is the large moment of the friction of rotation about the azimuthal axis due to the long arm of the friction force (the radius of the rolling circle).

In the rotatable supporting structures having a horizontal base there are also other specific disadvantages. For example, in the rotatable structures with turntables on carriages, the number of which is usually limited to four, or four groups, great concentrations of loading occur on the annular foundation under the circular track due to wind pressure [10]. This makes it necessary to build appropriate strength and rigidity into the foundation, which is then more time-consuming and expensive. This disadvantage exists also in the rotatable structures in which balls are used instead of carriages. However, here there is another drawback: the high cost of a ball train with balls of large diameter due to their complex technology with the use of unique machine tools.

2.1. Azimuthal Rotatable Supporting Structures

Based on Floating Circular Pontoons

To obviate the above-mentioned drawbacks of the conventional structural designs of azimuthal rotatable supporting structures for radio telescopes, we have proposed and developed a new structural design for support and azimuthal rotation in the form of a circular pontoon floating on a liquid in a circular basin or pool (see Fig. 3). In southern regions where the air temperature in the winter does not go lower than minus 10-15°C, the liquid in the pool may be an aqueous solution of sodium chloride.

The ring-shaped retaining wall of the circular pool may be made of concrete with a small percentage of reinforcing steel, since it is only slightly loaded. It serves only as a framing of the pool, protecting it from soiling. Its depth is determined by the depth to which the ground freezes. In southern regions the depth of the retaining wall (in the absence of seismic activity) may be about 1 m with a pool diameter of 50-100 m. In southern regions the bottom of the pool is simply leveled ground covered with a layer of loam, which prevents the filtration of the water.

The height of the liquid layer over the bottom (the depth of the pool), as will be shown below, does not exceed 1 m, which creates a pressure of about 0.1 kg/cm^2 on the ground, that is, from one-tenth to one-twentieth of the permissible pressure on the ground.

Centering stub 4 serves for centering the pontoon in the pool and for absorbing lateral forces from wind loads. Central bearing 14 permits the pontoon, without strain, to oscillate slightly with respect to the horizon.

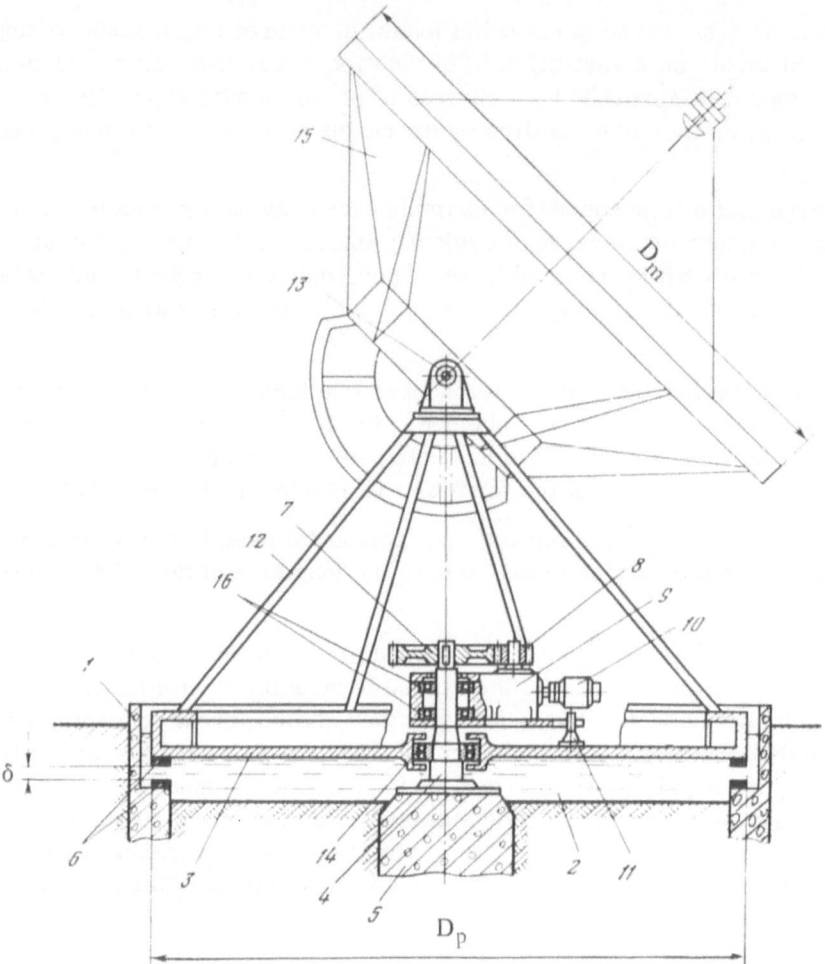

Fig. 3. Structural design (general view) of azimuthal rotatable supporting structure for the radio telescope, based on a floating circular pontoon (platform). 1) Reinforced-concrete retaining wall of
the circular pool; 2) the liquid in the pool; 3) circular pontoon (rotatable platform); 4) centering stub (column); 5) foundation of the
centering stub; 6) ring-shaped cover plates; 7) central (solar) gear
firmly attached to 4; 8) small driving gear on the output spindle
of a reducing gear-train unit; 9) the reducing gear-train unit; 10)
electric motor; 11) tangential-thrust stud; 12) supporting structure
of mirror system; 13) supporting journal with supporting bearing
of the elevation-angle axis; 14) central spherical roller bearing; 15)
main parabolic mirror; 16) bearings of the reducing unit's frame.

    Under the action of wind loads the floating pontoon will oscillate constantly. In order to
limit the amplitude of these oscillations, the rotatable supporting structure is provided with
ring-shaped cover plates 6 attached to the underside of the pontoon, at its periphery, and to
the ring-shaped pad of reinforced-concrete retaining wall 1. The amplitude of the pontoon's
oscillations is determined by the clearance or gap δ between the annular cover plates; this gap,
in turn, is determined by the amount of liquid in the pool. For filling and emptying the pool
there is a special reserve cistern with a pumping system, and for maintaining a certain level
of the liquid there is the usual float-switch control device.

Since the rotational resistance of the floating pontoon is about an order of magnitude less than (about one-tenth of) that of the conventional turntable on wheels (or on rollers), it follows that all the elements of the drive mechanism can be made substantially lighter. Moreover, the drive mechanism is simplified because of the absence of large-size machined parts. Thus, for example, one of the principal parts of the mechanism is the large gear; in the present design it can be comparatively small ($\sim 0.1 D_p$, where D_p is the diameter of the pool).

In the design we are describing, the entire drive mechanism (reducing gear train and electric motor) is mounted on one frame which is suspended by means of two bearings vertically spaced on the centering stub so that the rotating mechanism, together with the pontoon, is planetary (the large central gear being immobile). The frame of the drive mechanism is connected with the pontoon by tangential-thrust stud 11, which absorbs only a tangential force, not impeding the small radial and vertical displacements of the frame relative to the pontoon. These displacements occur when the pontoon oscillates. In spite of the pontoon's constant oscillatory movements, the accuracy of the gear meshing will not be disrupted, since the large gear and the small driving gear or pinion (at the output of the reducing gear-train unit of the azimuthal-drive mechanism) are strictly coupled together by their single (common) base, the centering stub. Therefore, the amount of clearance δ between the upper and lower annular cover plates is determined only by the sharpness of the directional diagram (the antenna's beamwidth).

The proposed rotatable supporting structure is calculated for use with radio telescopes using decimeter and meter wavelengths.

Let us examine, as an example, a radio telescope with a 100-meter parabolic mirror. Used for a wavelength $\lambda = 20$ cm, the width of its directional diagram (to the half-power level) will be

$$\theta_{0.5} = 68\lambda/D_m \simeq 8'.$$

For this directional diagram, the aiming accuracy of the radio telescope's antenna should not permit more error than $0.2\ \theta_{0.5}$, that is, about 1.6'.

According to preliminary calculations, the diameter D_p of the round pontoon is ~ 110 m (see below). With a pontoon oscillation amplitude $\varphi \approx 1'$, the linear size of the gap between the upper and the lower annular cover plates can be determined by the relation $\delta \leq D_p \tan \varphi$, and it is found to be ~ 34 mm.

The pontoon may be made as a rough metal-construction job without machining of the annular cover plates, the upper one of which is attached to the bottom (against the ring-shaped peripheral cylindrical wall) by means of bolts and washers, or spacers, for leveling. This way of mounting the cover plate can be done with an accuracy within ± 5 mm of deviation from the plane. The lower annular cover plate, which is attached to the ring-shaped reinforced concrete pad with anchor bolts, can be leveled without special labor to an accuracy within ± 5 mm with respect to the horizon.

The thickness of the liquid layer from the ground to the bottom of the floating pontoon is determined by the accuracy of the ground leveling, which, also without special labor, can be done with errors of not more than ± 100 mm. Consequently, it suffices to adopt a liquid-layer thickness of ≈ 200 mm.

If the radio telescope is to be located in northern regions, then the liquid should be nonfreezing down to temperatures in the -30 to $-40°$ range (for example, kerosene).

In this case, from the economic point of view, of considerable importance is the volume of liquid in the pool. It may prove more economical to increase the accuracy of the ground leveling and then by reducing the thickness of the liquid layer, reduce the volume of the liquid.

For determining the total depth of the pool, let us make an estimate of the required lifting force of the pontoon and the necessary pressure of the liquid on its bottom.

The total weight of the mobile (in azimuth) part of a radio telescope with its main mirror 100 m in diameter is determined by the relationship [11]:

$$G_{\phi 100} = G_{\phi 50}\,(D_{\phi 100}/D_{\phi 50})^{2.5}, \tag{12}$$

where $G_{\phi 100}$ and $G_{\phi 50}$ are the weights of 100- and 50-meter radiotelescopes, and $D_{\phi 100}$ and $D_{\phi 50}$ are the diameters of their main mirrors.

From the project of a radio telescope with a 50-meter main mirror, intended for operation in the decimeter-wave range, it is known that $G_{\phi 50} \approx 230$ tons; then from Eq. (12) $G_{\phi 100} \approx 1300$ tons. Consequently, for a 100-meter radio telescope, according to this estimate, the lifting force of the floating pontoon has to be not less than 1300 tons.

With an area of the pontoon's bottom of 9500 m^2 (its diameter is 110 m), the unit pressure on its bottom has to be ~140 kg/m^2, which corresponds to 140 mm of water column. Thus, the total height of the liquid's level (for water) over the floor of the pool is 20 + 14 = 34 cm, and the total depth of the pool is about 50–60 cm.

It was pointed out above that the radio telescope's moment of resistance to rotation in azimuth is, in the proposed design, approximately an order of magnitude less than in the conventional rotatable structures. Let us make a comparison of the moments of resistance to rotation.

The Hydraulic Resistance to Rotation of a Circular Pontoon (Floating Rotatable Platform). From Fig. 4 it is evident that the moment element of hydraulic-friction resistance is

$$\Delta M_{\mathrm{fr}}^{\mathrm{hydr}} = \Delta Q_{\mathrm{fr}}\,r, \tag{13}$$

where ΔQ_{fr} is a friction-force element acting on an annular area of radius r and width Δr. It is determined by the relationship

$$\Delta Q_{\mathrm{fr}} = 2\pi r \Delta r q_r C_{\mathrm{fr}}. \tag{14}$$

Here $q_r = \rho V_2^2/2$ is the velocity head (in kg/m^2); $C_{\mathrm{fr}} = f(R_e)$ is the coefficient of internal friction of adjacent layers of the liquid, being a function of the Reynolds number, $Re = L_r V_r/\nu$; $L_r = 2\pi r$ is the characteristic length (in meters); V_r is the linear velocity (in m/sec) directed along a tangent to a circumference of radius r, $V_r = 2\pi r n_s$. Usually $n_s = 0.0021$ rev/sec and $V_r = 0.0132r$; $\nu = 1.79 \cdot 10^{-6}$ m^2/sec is the coefficient of kinematic viscosity of water [12]. The

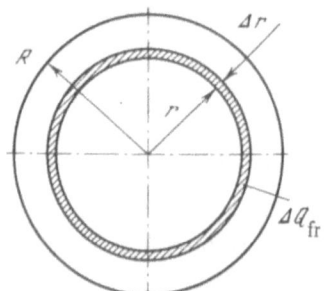

Fig. 4. Diagram for determining the hydrostatic moment of resistance to rotation.

mass density for water $\rho \approx 102$ kg$\cdot$sec^2/m^4 and $q_r \approx 0.0089r^2$ kg/m^2. Using the above numerical values for the Reynolds number, we derive

$$Re \approx 46,300r^2. \tag{15}$$

Since for small values of r the moment of the friction scarcely affects the total moment of friction, we shall make the estimate of Re for $r \geq 10$ m: $Re \approx 4.63 \cdot 10^6$. In this case the coefficient of friction in Eq. (14) is determined by a relationship from [12]

$$C_{fr} = 0.456/(\log Re)^{2.58} \tag{16}$$

or after substitution of the computed value of Re we get $C_{fr} \approx 0.00342$. After substituting the values in Eqs. (13) and (14), we have

$$\Delta Q_{fr} = 1.89\cdot10^{-4}r^3\Delta r, \quad \Delta M_{fr}^{hydr} = 1.89\cdot10^{-4}r^4\Delta r.$$

By integrating ΔM_{fr}^{hydr} with respect to r, we derive, for the moment of hydraulic-friction resistance, the equation

$$M_{fr}^{hydr} = \int_0^R 1.89 \cdot 10^{-4}r^4 dr = 0.378 \cdot 10^{-4}R^5. \tag{17}$$

With 2R = 110 m, $M_{ff}^{hydr} \approx 1.9 \cdot 10^6$ kg$\cdot$cm.

For comparison, let us make an estimate of the magnitude of the radio telescope's moment of resistance to rotation in azimuth in the case of a rotatable supporting structure on rollers. In this case the moment of resistance to rotation in azimuth is determined by a relationship from [13] (simplified)

$$M_{roll} = kfG_\Sigma R_{r.c}/r_r \quad \text{kg} \cdot \text{cm}. \tag{18}$$

where k is a coefficient that takes into account the sliding friction on the rolling circle $R_{r.c}$ and the friction in the bearings of the roller's axles, assuming it to be equal to 1.3; $f \approx 0.05$ cm is the coefficient of friction of the rollers' rolling; $G_\Sigma = 1.3 \cdot 10^6$ kg is the total weight of the radio telescope's mobile portion; $R_{r.c}$ and r_r are, respectively, the radius of the rolling circle (of the circular track) and the radius of the rollers; in our case $R_{r.c} = 55$ m and $r_r = 0.8$ m. After substituting these values in Eq. (18) we get $M_{roll} = 11.6 \cdot 10^6$ kg$\cdot$cm.

Thus, the moment of the resistance to the antenna's rotation when using a floating pontoon is almost an order of magnitude less than when using the customary rollers.

The effects of wind loads on the antenna, with the floating rotatable structure, reduce mainly to two factors (from the point of view of the rotatable structure's operation and that of the drive mechanisms): firstly, there results an oscillation of the floating pontoon, and secondly, there appears an additional frictional resistance whenever the annular cover plates (the one on the pontoon's bottom and the one on the pool wall's bottom) touch, which puts an additional load on the drive mechanism. Besides the additional moment of frictional resistance, generally there appears also an aerodynamic moment on the main mirror.

The moment of the wind load is determined by the following relationship:

$$M_V = M_V^m + Q_V h, \tag{19}$$

where M_V^m is the aerodynamic moment caused by an unequal distribution of wind-load pressures on different parts of the mirror; Q_V is the total force of the wind on the mirror, and h is its arm of application. In our case, where the mirror is of wire mesh, $M_V^m \approx 0$ and so

$$M_V = Q_V h.$$

For example, for a mirror 100 m in diameter, h ≈ 50 m and

$$Q_V = C_x \rho \frac{V^2}{2} \Sigma S, \tag{20}$$

where $\Sigma S = \Sigma l_i d_i$; l_i and d_i are the lengths and diameters of the rod (or wire) elements of the mirror structure. From preliminary computation $\Sigma S \approx 3100$ m²; $C_x = 1.4$ is the coefficient of aerodynamic resistance; and ρ is the mass density of the air.

When operating with a wind velocity of $V_{op} \sim 16$ m sec we have $Q_V \approx 6.9 \cdot 10^4$ and $M_{V_{op}} = 3.5 \cdot 10^6$ kg·cm.

Restoring Hydrostatic Moment. The pressure of the liquid on the pontoon's bottom is proportional to the depth of its immersion, since the pressure q varies linearly as is shown in Fig. 5a.

The pressure of the liquid on a bottom-side surface element which is at a distance x from the right-hand edge is

$$\Delta P = \Delta x b_x q_x,$$

where $b_x = (2Rx - x^2)^{1/2}$ and $q_x = qx/2R$.

The total pressure on the bottom is equal to the total weight of the overall floating-antenna unit

$$P = G_\Sigma = \frac{q}{R} \int_0^{2R} x \sqrt{ax^2 + bx} \, dx, \tag{21}$$

where $a = -1$ and $b = 2R$.

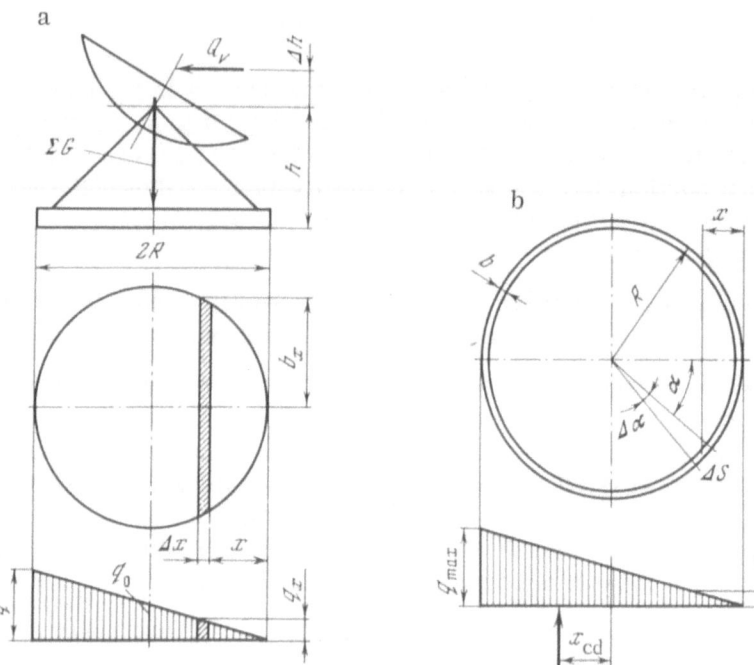

Fig. 5. Diagram for determining the additional hydrostatic moment of resistance to rotation, caused by wind loads. a) floating circular pontoon; b) toroidal pontoon.

After integrating Eq. (21) we have

$$\frac{R}{q} P = \frac{(ax^2 + bx)^{3/2}}{3a} - \frac{b}{2a}\left[\frac{(2ax+b)(ax^2+bx)^{1/2}}{4a} + \frac{1}{2k}\left(\frac{-1}{\sqrt{-a}}\arcsin\frac{2ax+b}{\sqrt{-\Delta}}\right)\right]\Big|_0^{2R}, \tag{22}$$

where $k = 4a/\Delta$, $\Delta = -b^2$.

After the pertinent substitutions and transformations we have

$$P = 0.5q\pi R^2. \tag{23}$$

With uniform distribution of the pressure of the liquid on the bottom ($q_{av} = G_\Sigma/\pi R^2$) and an inclined position of the pontoon (see Fig. 5a), the maximum pressure is

$$q = 2G_\Sigma/\pi R^2,$$

that is, double the average. For our example this comes to

$$q \approx 274.4 \ \text{kg/m}^2.$$

If the liquid is water, then the depth of immersion of the left edge of the pontoon is $h_{left}^{max} = 274.4$ mm, the right edge of the pontoon being at the water's surface ($q_{min} = 0$).

The hydrostatic moment of the pressure of the liquid on the pontoon with respect to the right edge is determined by the relationship

$$M_{hst} = \frac{q}{R}\int_0^{2R} x^2\sqrt{ax^2+bx}\,dx, \tag{24}$$

where a and b have the same values as in Eq. (21).

After integrating Eq. (24) we have

$$M_{hst} = \frac{q}{R}\left\{\left(x - \frac{5b}{6a}\right)\frac{(ax^2+bx)^{3/2}}{4a} + \frac{5B^2}{6a^2}\left[\frac{(2ax+b)(ax^2+bx)^{1/2}}{4a} + \frac{1}{2k}\left(\frac{-1}{\sqrt{-a}}\arcsin\frac{2ax+b}{\sqrt{-\Delta}}\right)\right]\right\}\Big|_0^{2R} \tag{25}$$

[here k and Δ are the same as in Eq. (22)]. By substituting the pertinent values and transforming, we get

$$M_{hst} = {}^5/_8\pi qR^3. \tag{26}$$

The distance of the resultant force of hydrostatic pressure from the right edge of the pontoon is determined by the relationship

$$x_P = M_{hst}/P = 1.25R.$$

The restoring hydrostatic moment with respect to the center of the pontoon is

$$M_{hst}^r = P(x_P - R) = 0.25PR,$$

and in the specific case it is $M_{hst}^r = 18 \cdot 10^6$ kg·cm.

The angular amplitude of the pontoon's oscillations and, consequently, of the whole antenna is

$$\Delta\varphi = h_{left}^{max}/2R \approx 0.0025 \ \text{rad} \approx 8.6'.$$

It has been mentioned above that the proposed floating azimuth-rotatable supporting structure is intended for a radio telescope of decimeter and longer wavelengths. With a wavelength λ = 20 cm, the width of the directional diagram is $\theta_{0.5}$ = 8'. The angular amplitude of the pontoon's (and antenna's) oscillations is $\Delta\varphi \approx 8.6'$, that is, it exceeds the value of $\theta_{0.5}$. But this situation occurs at maximum tilt of the antenna, when the right edge of the pontoon (platform) rises to the surface of the liquid, but still does not rise above it. For these conditions the restoring hydrostatic moment was calculated from the wind loads resulting from a wind velocity considerably greater than our operational V_{op} = 16 m/sec, in which case it can be verified by using the relationship between wind velocity and maximum hydrostatic pressure.

$$ V = R \sqrt{\frac{2q\pi R}{C_x h \Sigma S}} . \tag{27} $$

In our case, with q = 274.4 kg/m^2, we have C_x = 1.4, h = 50 m, ΣS = 3100 m^3, R = 55 m, and the wind velocity V $\approx$ 36.5 m/sec, that is, almost a hurricane. Since we are assuming the operational wind velocity to be $V_{op} \approx$ 16 m/sec, the maximum operational hydrostatic pressure from wind loading [see Eq. (27)] will be considerably lower: q $\approx$ 54 kg/m^2.

The angular amplitude of the pontoon's (and antenna's) oscillations is

$$ \Delta\varphi = \Delta h_{max}/2R \simeq 0.00049 \text{ rad} \simeq 1.7'. $$

Thus, the variation in the antenna's direction, due to oscillations of the pontoon, is almost five times smaller than the width of the directional diagram.

2.2. Azimuthal Rotatable Supporting Structure

Based on a Floating Toroidal Pontoon

The above-examined azimuthal rotatable supporting structure based on a floating circular pontoon has the disadvantage that in the northern regions the bottom of the pool must be sufficiently strong and rigid so that it can preclude any bulging from the ground's freezing and thawing. This need involves great difficulty and high cost in the preparation of the ground and also the cost of a large volume of liquid, which in such a climate will not be water, but, for example, transformer oil. In order to obviate this disadvantage, we have proposed and developed a variant of the floating pontoon: one in the form of a torus to be placed in a ring-shaped canal filled with a nonfreezing liquid.

A ring-shaped canal with a semicircular cross section can rest on piles and be situated at some distance from the ground. Naturally, the diameter (of the midline) of such a ring-shaped canal, as also that of the toroidal pontoon, must be greater than the diameter of the circular pool and of the corresponding pontoon. However, the increase of pontoon diameter has its advantage, since it makes possible an increase in the number of supports that transmit the load from the mirror structure to the rotatable platform, that is, to the torus, without limitation of rotations of the main mirror and with a relatively small distance of it from the horizontal axis of rotation.

For a toroidal pontoon we have the following relationships (see Fig. 5b):

$$ M_{hst} = bR^2 q_{max} \int_0^{\pi} (1 - \cos\alpha)^2 \, d\alpha, $$
$$ P = 2\pi R b q_{av}, \quad q_{max} = 2q_{av}, \tag{28} $$

and

$$ q_{max} = \frac{C_x \rho V^2 h \Sigma S}{2b\pi R (x_{c.p.} - R)} . \tag{29} $$

With R = 75 m, ΣS = 3100 m^2, h = 50 m, b = 3 m, and C_x = 1.4, we have q_{max} = 131.2 kg/m^2; x is the arm of the center of pressure.

The angular amplitude of the oscillations of an antenna that is on a toroidal pontoon is (when using as the liquid an oil of γ = 0.9 ton/m^3)

$$\Delta\varphi = q_{av}/2R\gamma \approx 0.001 \text{ rad} \approx 3.6'.$$

A preliminary structural study shows that a floating toroidal pontoon used for the azimuthal rotatable supporting structure of a radio-telescope antenna is almost independent of the climate, requires a smaller amount of metal, and is more economical (however, the final selection of the type of floating pontoon should be made on the basis of more detailed structural studies and calculations with allowance for the specific climatic conditions of the site where the radio telescope is going to be erected). See Fig. 6.

2.3. Azimuthal Rotatable Supporting Structure

Based on an Air Cushion

At the P. N. Lebedev Physics Institute (in the Radio Astronomy Laboratory) there has been proposed and developed a new structural design of an azimuthal rotatable supporting

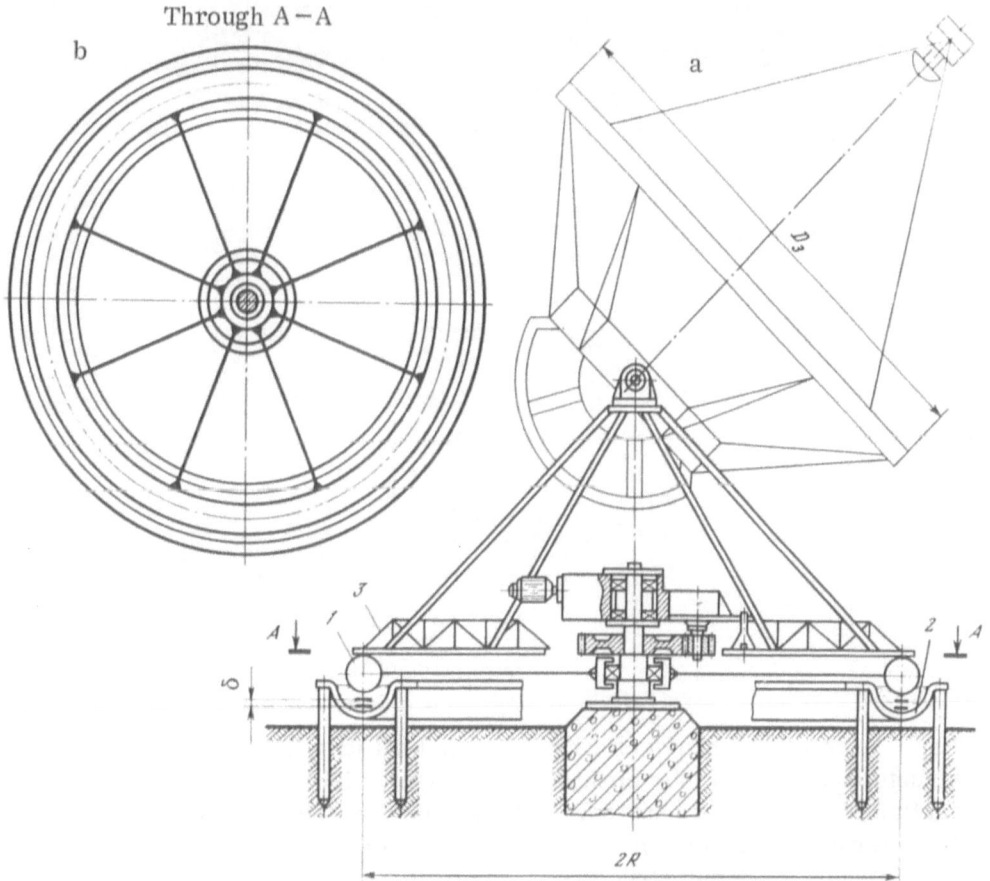

Fig. 6. Structural design of azimuthal rotatable supporting structure based on a toroidal pontoon (a) and its top view or plan through A−A (b). 1) Toroidal pontoon; 2) ring-shaped canal with nonfreezing liquid; 3) diametric bridge. The remaining units are the same as in Fig. 3.

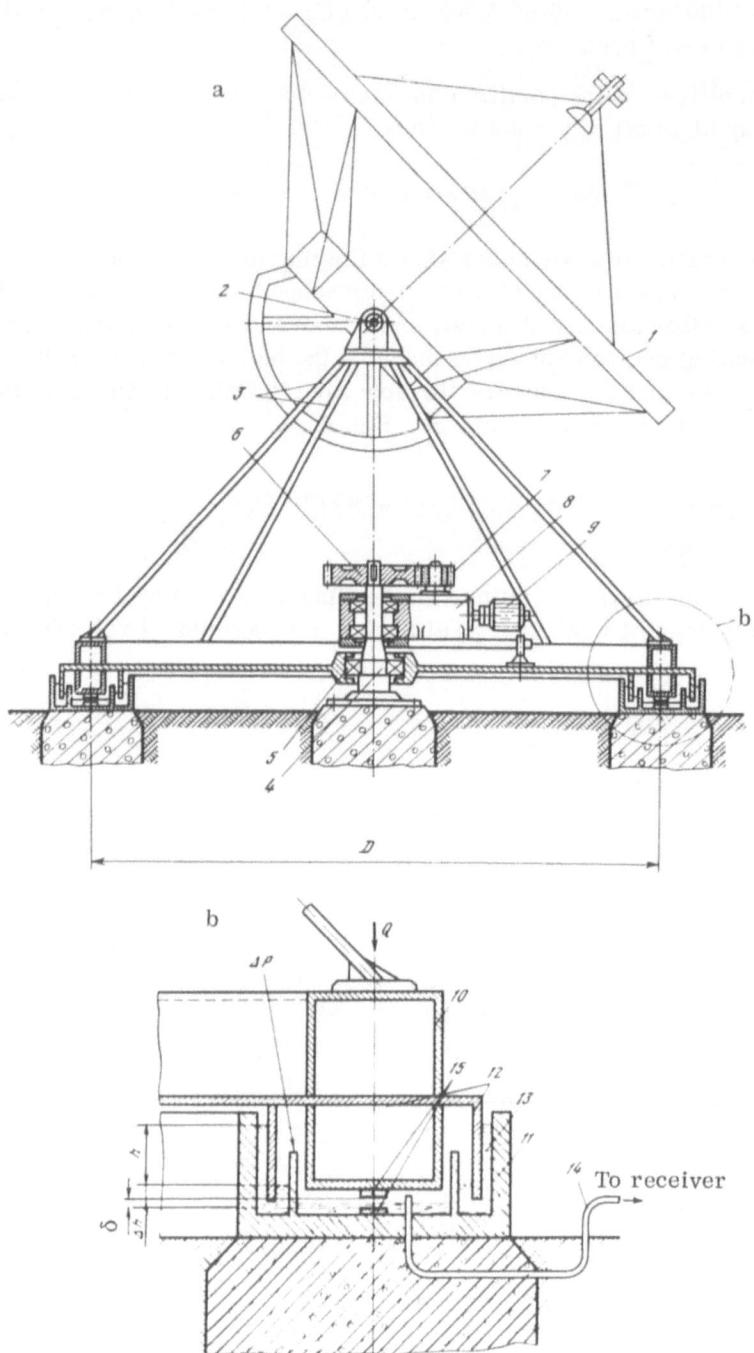

Fig. 7. Structural design of a rotatable radio-telescope supporting structure based on an air cushion. a) General view; b) local cross section of the toroidal pontoon, platform, and ring-shaped canal. 1) Main parabolic mirror; 2) supporting pins of the elevation-angle axis; 3) supporting bar structure; 4) centering stub; 5) central spherical ball bearing; 6) large (solar) gear; 7) driving gear; 8) reducing gear-train unit of the azimuth-drive mechanism; 9) electric motor; 10) annular platform; 11) annular canal; 12) annular walls of the lock; 13) nonfreezing liquid; 14) air duct; 15) annular cover plate.

structure of a radio telescope which is based on an air cushion — more precisely speaking, on a static air cushion, because, as usually understood, the air cushion used in transport or conveying equipment is an aerodynamic air cushion. Whereas the aerodynamic air cushion requires an enormous expenditure of air and a corresponding expenditure of energy, on the other hand the static air cushion is free from this defect. In this latter case the expenditure of air is insignificant.

This rotatable supporting structure based on an air cushion consists of the following basic units: (a) an annular platform with annular locking walls (Nos. 10 and 12 in Fig. 7b); (b) annular canal with annular locking walls; (c) centering stub; (d) azimuth-drive mechanism; (e) supporting structure of the mirror system.

Annular platform 10 is a welded steel-plate structure. The cross section of this annular structure is shaped as a closed rectangle whose lower part, subjected to high-pressure air ΔP, must be hermetic. The lower part of the annular platform is provided with annular walls 12 situated to the outer and inner sides of the annular platform (concentrically) and serving as locks that separate the region of high air pressure inside annular canal 11 from the atmospheric pressure, by means of liquid 13 (for example, oil). On the underside of the annular platform, along its midline, is attached annular cover plate 15 (of Teflon or bronze).

The centering of the annular platform and canal is accomplished by means of centering stub 4, to which the platform is bound through central spherical ball bearing 5, whose frame is fastened to the annular platform's diaphragm. This diaphragm (a horizontal plate or a system of spokes), which stiffens the platform in its plane, can transmit a transverse load from the mirror system to the centering stub.

The annular canal, having an open cross section (open upward), serves for placing the annular platform in the liquid, that is, for creating the static air cushion. The annular canal also contains interior ring-shaped walls which form the immobile part of the lock. These walls are concentric with the canal's outer walls. Along the midline of the annular canal (on its bottom) is attached an annular cover plate corresponding to the one on the underside of the annular platform. The annular canal may be made of monolithic reinforced concrete as a single unit together with its annular foundation.

The azimuth-drive mechanism consists of solar gear 6 mounted on the centering stub and reducing gear-train unit 8 whose frame is suspended on the centering stub by two ball bearings that are spaced vertically. The unit's frame rotates around the centering stub together with the annular platform to whose diaphragm it is connected by a stud.

Supporting structure 3, which connects the mirror system (together with its suspension) to the annular platform, is made in the form of two bar pyramids (corresponding to the two supporting pins 2) whose vertices are flat, and on these flat areas are situated the supporting bearings of the mirror system's elevation-angle axis. The lower ends of the supporting structure's bars are made fast to the periphery of the annular platform at equal distances along the arc of the ring, ensuring uniform loading on the annular platform from the weight being supported, and a linear distribution of any load from wind pressure.

The operation of a radio telescope's rotatable supporting structure based on an air cushion proceeds as follows: In the annular canal's inner space included between its bottom and the lower part of the annular platform, and isolated from the outer atmosphere by the annular lock with liquid, air at a pressure of about ≈ 1.10 kg/cm^2 is pumped in. The liquid of the locking system is displaced by the overpressure ΔP from the inner cavity of the locking system to the outer cavity. With the attainment, in the inner cavity of the annular canal's locking system, of an overpressure balancing the weight of the radio telescope's azimuthal-rotatable portion, the annular platform, which until then had been resting on the annular cover plates, is lifted to a height δ necessary for its free rotation in azimuth. The amount of clearance δ is selected in accordance with the permissible amplitude of the antenna's oscillations from wind loading, and this permissible amplitude, in turn, is determined by the width of the directional diagram.

With the maintenance of weight centering, with which the center of gravity matches the center of the annular platform, and in the absence of wind, the radio telescope's moment of resistance to rotation in azimuth is determined by the hydrodynamic resistance of the locking system. The moment of resistance to rotation in azimuth is determined by the relationship

$$M_{hst} = C_{fr} \, \Sigma S \rho \, \frac{V^2}{2} \, R. \tag{30}$$

For concreteness let us use the example considered above. In this example we consider a radio telescope with a fully rotatable and inclinable parabolic antenna of 100-m main-mirror diameter. The weight of the antenna's rotatable part is G_Σ = 1300 tons; the diameter of the annular platform is 2R = 150 m; and the cross section of the annular platform, or torus, we take to be of rectangular shape: $ab = 2.5 \times 3.0 = 7.5$ m^2 (in the example given before, the cross section of the torus was a circle 3 m in diameter).

The depth of immersion of the annular platform into the annular canal (filled with oil) will be

$$h = G_\Sigma/(2\pi R b \gamma), \tag{31}$$

where $2\pi R$ = 471 m, b = 3 m, γ = 0.9 ton/m^3, and the result is h ≈ 1.02 m.

The moment of resistance to rotation in azimuth is determined from the moment of the friction forces between the annular walls of the canal (inner surfaces) and the annular walls of the platform's lock, and the liquid that fills the annular canal:

$$M_{hd} = Q_{hd} \, R, \tag{32}$$

where $Q_{hd} \approx C_{fr} \rho V^2 \Sigma S/2$; $\Sigma S = 2(2\pi Rh') = 1300$ m^2 is the total surface of friction; h' = h + Δh = 1.35 m.

For oil with a specific weight γ = 0.9 ton/m^3, the mass density is $\rho = \gamma/q = 81.5$ kg $\cdot$ sec^2/m^4.

Let us assume a linear tangential velocity, for the pontoon, of V = 0.7 m/sec. The coefficient of friction (between the inner layers of the oil) is found from Eq. (16) using the formula for the Reynolds number Re = LV/ν and applying the value of the kinematic viscosity coefficient for transformer oil at a temperature t = 30°C, which is $\nu_{t=30°} = 0.3 \cdot 10^{-4}$ m^2/sec; $C_{fr} \approx 0.0044$.

Thus, the hydrodynamic resistance from friction is

$$Q_{hd} \approx 114 \, kg$$

and the moment of resistance to rotation resulting from Q_{hd} is

$$M_{hd} \approx 0.82 \cdot 10^6 \quad kg \cdot cm.$$

At low temperatures the coefficient of friction C_{fr} increases in accordance with the increase of the kinematic coefficient of viscosity ν. According to data in [14], for some aviation oils and for low temperatures the temperature dependence of ν is determined by the relationship

$$\nu = \nu_{t=30°} \left(\frac{75}{t+40} \right)^m,$$

where m = 0.49-0.79. Assuming m = 5, we derive the value at t = −25°: $\nu_{t\,=\,-25°} \approx 1.0$ m²/sec. With this, the Reynolds number will be 348. For this value of Re the coefficient of friction is determined by the relationship

$$C_{fr} = 1.328/\sqrt{Re},$$

that is, C_{fr} = 0.0711. The resistance from friction in this case is $Q_{hd} \approx 1850$ kg and the moment of resistance $M_{hd} \approx 1.4 \cdot 10^6$ kg·cm.

For a rotatable supporting structure on rollers, the moment of resistance to the rotation of the antenna in azimuth [see Eq. (18)], with the platform's rolling-circle radius R = 75 m, the roller radius r = 0.4 m, and k = 1.5, is $M_{roll} \approx 9.8 \cdot 10^6$ kg·cm.

Thus, even for the low temperature t = −25°C, the moment of resistance to the rotation of a rotatable structure in azimuth on an air cushion is about one-seventh of that for the conventional rotatable structure based on rollers.

In the analysis of the effect of wind load, preliminary calculations have shown that even for a moderate breeze of velocity V = 5 m/sec, the additional moment of resistance to rotation is $\Delta M = 5.4 \cdot 10^6$ kg·cm, that is, approximately 3.5 times greater than the moment of hydrodynamic resistance.

Consequently, only for the limiting conditions, that is, in the absence of wind loads and with centering of the weight, can the rotatable structure based on an air cushion provide a substantial advantage in the value of its moment of resistance to the rotation of the radio telescope in azimuth and a corresponding advantage in the weight of the drive mechanism's power units and of the rotatable structure's carriers.

It would be most appropriate to apply the proposed design of rotatable supporting structure to radio telescopes that are inside shelters.

2.4. Elevation-Angle Rotatable Structure

One of the problems which must be solved in the designing and building of the fully rotatable and inclinable parabolic antennas of the large radio telescopes is the problem of ensuring a rigid orientation of the radio telescope's mirror system. A slight mobility, that is, an insufficient rigidity, of the radio telescope's antenna manifests itself in the form of angular displacements with respect to the axes of the rotatable and inclinable supporting structure when there are accidental disturbances, for example, from gusts of wind. These small angular displacements of the radio telescope's antenna are possible because of the existence of play between kinematic couples of the drive mechanisms and of elastic play in the couples themselves. The magnitude of these displacements is inversely proportional to the size of the couples' components. For example, the greater the diameter of the toothed wheel or gear mounted on the natural axis of a rotatable structure, the less is the corresponding elastic play (in angular measure). Therefore the most radical means of decreasing angular play, that is, of increasing the rigidity of the elevation-angle orientation of the radio telescope's antenna—mirror system, is to increase the diameter of the toothed (or bobbin) sector of the elevation-angle drive mechanism.

So long as radio telescopes were comparatively small, for example with a main-mirror diameter of 5 to 10, the fabrication of the toothed sector, whose diameter would be comparable to that of the main mirror, was not a problem. But when radio telescopes attained a main-mirror diameter of 22-27 m, and a diameter of the large toothed sector of the order of 12-15 m, some difficulties developed in their fabrication. In the fabrication of the RT-22 radio telescopes of the Physics Institute of the Academy of Sciences of the USSR and of the Crimean Astrophysical Observatory [15, 16], these difficulties were surmounted by having the drive sector made as a bobbin instead of a gear.

In the construction of radio telescopes having a main-mirror diameter of ≈26 m for the National Radio Astronomic Observatory (NRAO) in the United States of America, the large toothed sector with a diameter of about 13 m was made as a composite toothed crown of individual sections of toothed rack bent over the cylindrical surface of the sector's rim.

In the 100-meter radio telescope of Bonn University (Federal Republic of Germany), the toothed crown of the large toothed sector, the diameter of whose initial circumference is 56 m, was also made from straight toothed racks that were curved in a press to arcs of the desired circumference. In the process of meshing, the intercenter distance does not remain constant, and the continuity of the meshing is achieved by maintaining a constant distance between the teeth of the racks and those of the small gear, or pinion. In order to keep this distance constant, a carriage for supporting the reducing-gear train of the drive mechanism is clamped to the rim of the system of guide rollers [10].

Thus, the fabrication of large gears (or toothed sectors) involves well-known difficulties. Furthermore, the construction of the connection of the toothed sector (of its hub) to the suspension of the main mirror is also rather complicated.

Further increase in the size of radio telescopes (and the corresponding increase in the size of the toothed sector) leads to a still greater increase in the difficulties of its fabrication and mounting.

In recent years there have appeared fundamentally new structural designs of elevation-angle tilting mechanisms. For example, in England there has been patented an elevation-angle drive mechanism that has, as a basic unit, a screw couple in which the advance of a nut along the axis of a screw is transformed into a tilting of the radio telescope's mirror−antenna system in azimuth [17].

A structural design of elevation-angle tilting mechanism with two parallel-operating screw pairs has been applied in the project of a radio telescope with a main mirror of ≈32 m diameter developed by the Marconi firm (1958) [18]. It must be pointed out that the design with a screw pair has substantial advantages as compared to the conventional tilting mechanisms that comprise an open toothed drive.

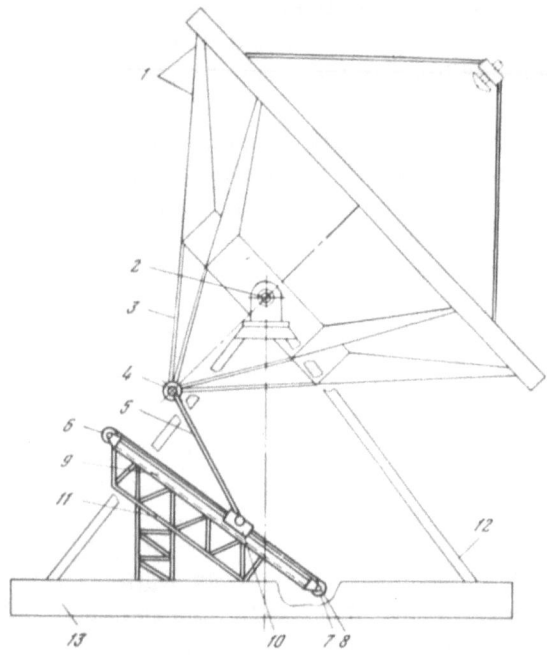

Fig. 8. Design of radio telescope's elevation-angle tilting mechanism with traction carriage. 1) Main inclinable parabolic mirror; 2) supporting pivot at the elevation-angle axis; 3) supporting bar pyramid; 4) vertex of the pyramid and upper hinge end of the connecting bar; 5) connecting bar; 6, 7) upper and lower controlling sprocket wheels; 8) drive shaft (spindle) of the reducing-gear unit (the reducing-gear unit of the drive mechanism is not shown); 9) traction chain; 10) traction carriage; 11) supporting truss of the trestle bridge; 12) a bar of the mirror system's supporting structure; 13) rotatable supporting platform.

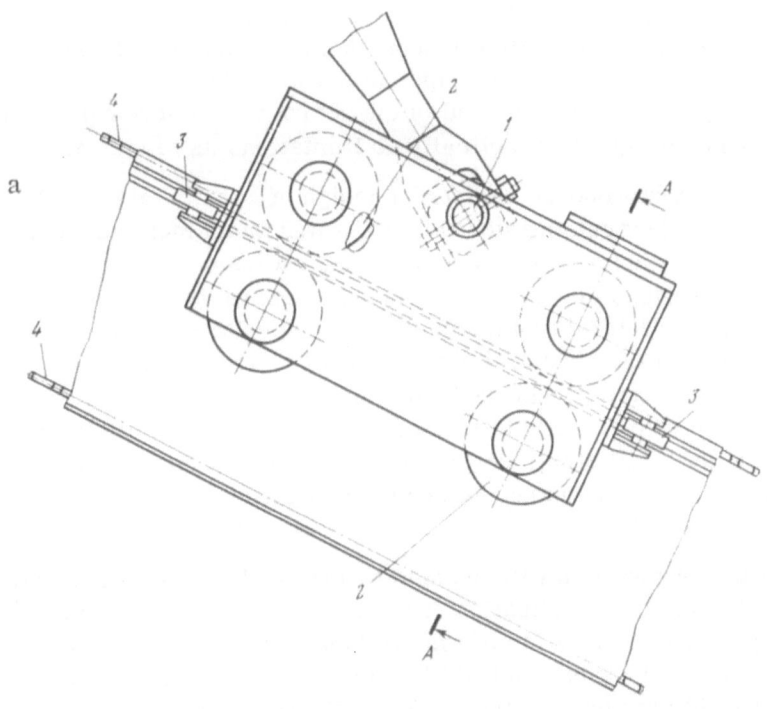

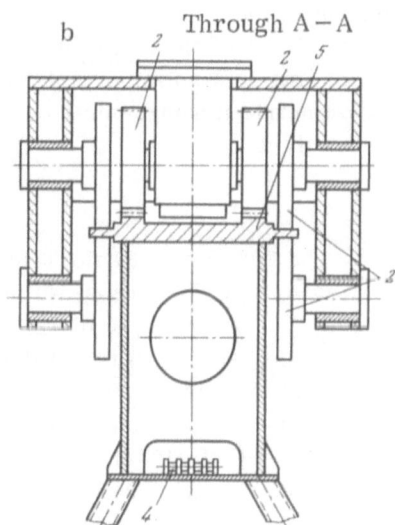

Fig. 9. Traction carriage. a) Side view; b) cross section; 1) pin of the connecting bar's lower hinge; 2) guide rollers; 3) lateral guide rollers; 4) traction chain; 5) guide strip.

The screw drive is simpler and easier to make, can be made without play, and weighs considerably less. However, it has its disadvantages, which limit the possibilities of using it for large radio telescopes. First, the size of the screw is limited by the possibilities of its fabrication by existing screw-cutting lathes (obviously, for it is about 15-20 m in length); and second, the size of the screw is limited by structural considerations: the low rigidity of the console or free (unsupported) end of the screw, which can lead to the appearance of vibrations.

We have proposed a new structural design for a radio telescope's elevation-angle drive mechanism, the basic unit of which is a traction carriage that is moved along a straight guide by means of a traction chain. In this design, just as in the screw pair, the carriage's advance is transformed into a tilting of the radio telescope's mirror-antenna system. However, the structural design proposed in the present paper is devoid of the disadvantage both of the conventional design with gear transmission and of the design with a screw pair. The present design with a traction carriage has practically no limitations as to size and is simple to make.

The basic units of the proposed structural design of a radio telescope's elevation-angle tilting mechanism are: (a) the guiding steel strip situated on an inclined flat trestle-bridge surface, the length of which is approximately equal to the radius of the main mirror (see Fig. 8); (b) the traction carriage with guide rollers ensuring inseparable and playless motion along the guide strip, and preventing lateral displacement and all possibility of jamming (Fig. 9); (c) the two sprocket wheels situated at the ends of the trestle bridge: the upper one is a guide and the lower one is for traction (drive); (d) the reducing-gear unit of the drive mechanism, mounted on one frame so that the spindle of the traction sprocket is connected with the reducing-gear unit by gear transmission (of small size); (e) the connecting bar, the ends of which connect by hinges: the lower end with the body of the carriage and the upper end with the supporting bar pyramid whose base is rigidly bound to the supporting framework of the main mirror.

In order that the proposed elevation-angle tilting mechanism operate without any slack or play, that is, permit an exact inclination of the radio telescope at the prescribed angle, it suffices to ensure the exact fabrication of the two hinges of articulation at the ends of the connecting bar and a constant thickness and width of the guide strip along which the traction carriage moves. A highly accurate flatness and straightness of the guide strip is not required.

The use of the above-described elevation-angle tilting mechanism is inherently related to the use of the multiple suspension of the main mirror with a radially symmetric arrangement of supports in the form of a supporting bar pyramid, which ensures radially symmetric deformations of the main mirror due to its own weight with any orientation of the radio telescope. This is also important, since the multiple suspension of the main mirror with the radially symmetric arrangements of supports ensures great rigidity of the mirror with a small structural weight.

LITERATURE CITED

1. The 100-meter Radio Telescope of the Max Planck Institute in Bonn. Collection of Translations, ONTI Press, No. 3 (1971), p. 22.
2. V. V. Vitkevich, Radio Interferometers and Radio Telescopes at the Crimean Station of the Physics Institute of the Academy of Sciences. Proceedings of the 5th Conference on Problems of Cosmogony, 1955 [in Russian], p. 14.
3. Laurenz Mohr, Über Konstruktion und Berechnung des antennen Spiegels, Der Stahlbau, lg. 27, H. 3 (1958).
4. V. V. Vitkevich and P. D. Kalachev, Tr. FIAN, 28:39 (1965).
5. P. D. Kalachev, Tr. FIAN, 62:150 (1972).
6. P. D. Kalachev, V. P. Nazarov, I. A. Emel'yanov, and V. L. Shubeko, Preprint FIAN, No. 51 (1973).
7. N. M. Belyaev, Strength of Materials [in Russian], Gostekhizdat, Moscow (1953), p. 114.
8. J. P. Wilde, Usp. Fiz. Nauk, 92(4):706 (1967).
9. O. Khakhenberg (Hahenberg), Collection of Translations, ONTI Press, No. 3 (1971).
10. E. Geldmacher, Tech. Mitt. Krupp, Werksber., 28(4) (1970).
11. B. A. Garf, Utilization of Solar Energy [in Russian], Izd. AN SSSR, Moscow (1957), p. 62.
12. Mechanical Engineering Handbook [in Russian], Mashgiz, Moscow (1950), Vol. 1, pp. 790-808.
13. A. G. Lang and V. S. Maizel', Harbor Cranes [in Russian], Mashgiz, Moscow (1953), p. 117.

14. Encyclopedic Handbook of Mechanical Engineering [in Russian] (1947), Vol. 1, Part 1, p. 385.

15. P. D. Kalachev and A. E. Salomonovich, Tr. FIAN, 17:13 (1962).

16. V. I. Ivanov, I. G. Moiseev, and Yu. G. Monin, Izv. Krym. Astrof. Observ., 38:141 (1967).

17. English patent No. 1,234,231, cl. H4A from 26 January 1968.

18. Equipment for Radio Astronomy Systems, Marconi Space Communication Systems, P-60-089G (1969).

PRECISION SYSTEM FOR GUIDING THE RT-22 RADIO TELESCOPE OF THE PHYSICS INSTITUTE OF THE ACADEMY OF SCIENCES OF THE USSR

V. A. Vvedenskii, P. D. Kalachev, A. D. Kuz'min, Yu. N. Semenov, and R. L. Sorochenko

A new automated system for guiding the RT-22 radio telescope of the Physics Institute of the Academy of Sciences of the USSR, operating from a universal M-6000 electronic computer, is described. The radio telescope's operating procedures, which are provided for by the new system, are examined. The design of an azimuth data-unit installation that ensures the system's operation with an accuracy of 10" is described.

One of the key problems in the creation of radio telescopes of great resolving power is the development and fabrication of precise systems for aiming and guiding them. For the RT-22 radio telescope of the Physics Institute of the Academy of Sciences of the USSR, the first such system was developed in 1954-1957 [1] and was in operation from 1960 through 1975.

The system was built with analog elements and provided for the automatic aiming of the radio telescope to a point with prescribed equatorial coordinates and for the tracking of this point with an error of 2-3 minutes of arc.

This procedure was employed for work with wavelengths longer than 5 cm, where the width of the directional diagram was greater than 10 minutes of arc. In the millimeter range the aiming and tracking were accomplished with a semiautomatic procedure of visual guidance for observing the object by means of a small optical telescope. This procedure provided the necessary tracking accuracy. However, it can be employed only when the object is visible, that is, only on a clear night. The duration of night averages about 40% of the 24-hour period, and the average number of clear nights in a year in the region of the Physics Institute's astronomical observatory is about 20%. Thus the radio telescope's useful time with the procedure of visual guidance was only 8%; that is, the effective utilization of this unique instrument, equipped with highly sensitive radiometers, was poor.

In order to increase its effectiveness, it was necessary to substantially enhance the accuracy of the automatic procedure of aiming and tracking in order to operate in the millimeter range without optical visibility of the object being studied. The required enhancement of accuracy could be achieved only by changing from the analog to a digital system, and this was done in the control system of the RT-22 radio telescope at the Crimean Astrophysical Observatory [2].

However, this system employed a specialized electronic computer with an inflexible program, which limited the observational possibilities of the radio telescope.

Theoretically, a control system working from a universal electronic computer (EC) has other possibilities. Such a system is being created at the present time for the RT-22 of the Physics Institute of the Academy of Sciences of the USSR in accordance with the project of comprehensive automation of radio-astronomic research with this instrument by means of an M-6000 electronic computer [3]. A block diagram of the system is presented in Fig. 1.

The M-6000, completable by the aggregation principle, is equipped with the input and output devices necessary for its inclusion in the radio telescope's circuits. The computer input data, the selection of programs, and the monitoring of the passage of observations and operating actions of the observer will be effected through a "Konsul-260" keyboard printer and an SID-1000 data indicating station (DI St).

The system employs units that fully satisfy modern technical requirements:

1. A PIUK-17 18-order digital data unit is used as a follow-up instrument.
2. Amplifying units of the tracking drives are made on the basis of microcircuits.
3. There is provision for digital indication of the prescribed angles in different coordinate systems, displayed on vacuum luminescent triodes with segmental control;
4. For dial monitoring of the telescope's position, an industrial television (IT) is provided in the system.
5. For the convenience of the operators, the system is equipped with loudspeaker-and-telephone communication.

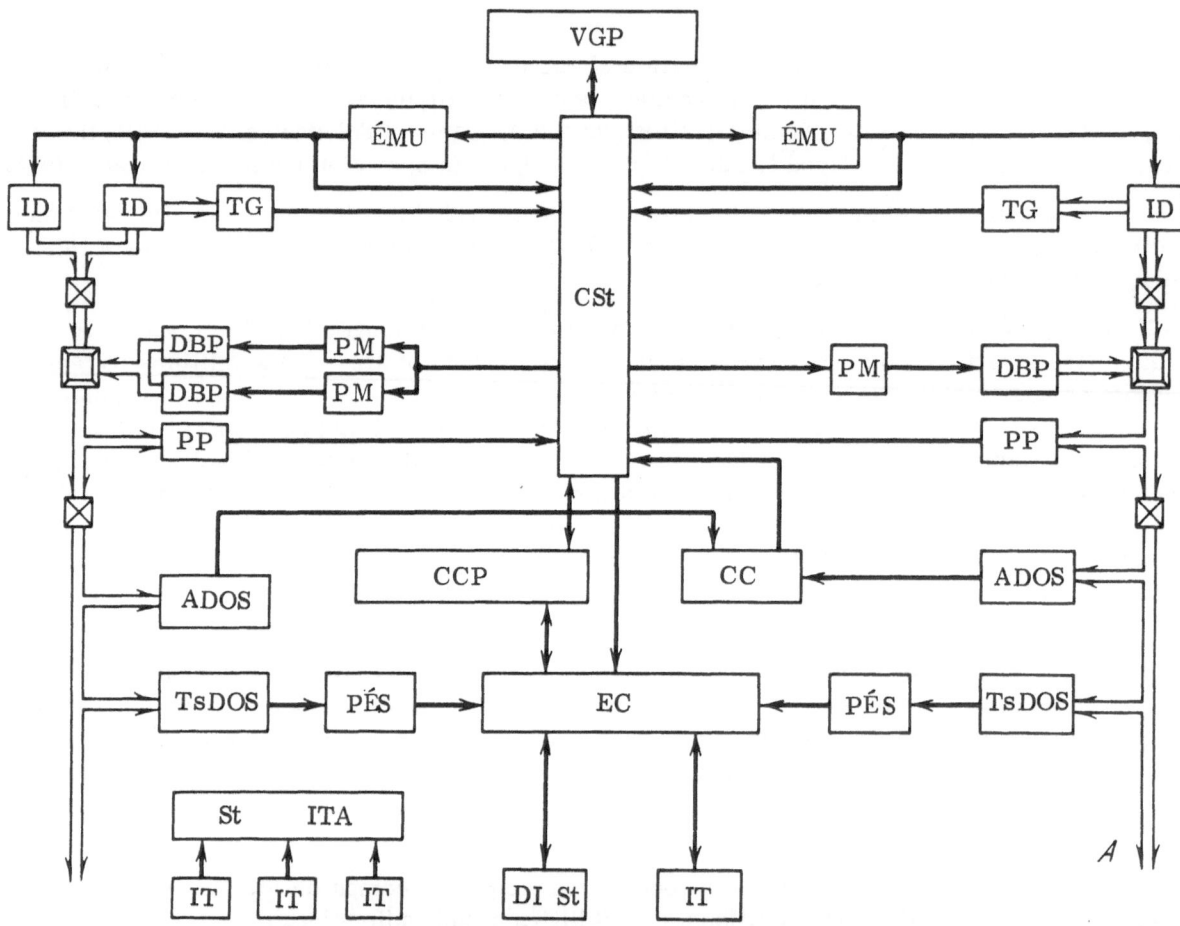

Fig. 1. Block diagram of the system.

Time correlation will be effected by the system of unique time and frequency of the radio-astronomic standards of the Physics Institute of the Academy of Sciences of the USSR [4].

On the basis of modern elements for the Physics Institute's RT-22 system, the following units have been redeveloped: the central control panel (CCP), the visual guidance panel (VGP), the control station (C St), and the station of industrial television apparatus (St ITA), involving instruments and devices, tachometer-generator units, angle stops (limiters), and new arrangements of data-unit and drive mechanism installation.

The CCP, the St ITA, and the electronic computer with equipment for lead-in and display are located in a laboratory house, while the rest of the system's apparatus will be placed on the radio telescope.

The new automatic tracking system for the Physics Institute's RT-22 will provide for the following operating procedures: (1) quick resetting; (2) automatic guidance from the electronic computer; (3) automatic guindance from the analog converter of coordinates (CC); (4) semiautomatic visual guidance with the help of an optical guide telescope.

The quick resetting of the radio telescope will be accomplished by nonsynchronous motors with open-circuit control.

The procedures of automatic tracking from the analog CC and the semiautomatic visual procedure in the new system are analogous to corresponding procedures in the old one.

Automatic operation from the electronic computer provides for (as chosen by the operator): (1) the aiming of the radio telescope at a prescribed point in space for observation by the transit method; (2) the tracking of an object having constant coordinates α and β; (3) the tracking of an object having varying coordinates α and β; (4) scanning in coordinates α and β; (5) scanning in azimuth and elevation angle; (6) alternate tracking of two sources.

For guidance by digital slave drives in the automatic procedure, the electronic computer will, twice per second, generate in digital form a signal of mismatch between the computed value of the angle and the real value being read out from a digital follow-up instrument in each of the coordinates: azimuth and angle of elevation. Besides the mismatch signal the computer will also generate, in digital form, a signal proportional to the rate of change of the computed coordinates (a signal of the prescribed rate of change).

The signals of mismatch and of prescribed rate of change have discreteness in level and time. A discrete signal of the prescribed rate, upon entering the input of a slave drive from the computer's converting systems, is compared to a continuous signal proportional to the velocity of the system's exit spindle (the signal of the taking velocity). The slave drive is then governed by a signal proportional to the difference between the prescribed and the taking velocities. In velocity control an accumulation of angular error occurs. If this error exceeds $5''$, then, with the next exchange of information, an angular-correction signal goes to the slave drive from the converting systems of the electronic computer. In this way, the slave drive of each axis is controlled from the computer as to velocity and with correction of angle (combined guidance).

Laboratory tests of the system have shown that, with the automatic operating procedure from the electronic computer, it is possible to ensure tracking with an error of 1-2 units of discreteness of level, that is, $5-10''$.

The rotatable supporting structure of the radio telescope allows lateral displacements of the rotatable platform in different directions in the amount of the radial clearance of 0.4-0.5 mm between the rotating and stationary parts. This leads to error of azimuth angle-of-turn readout. The usual mounting of the data-unit mechanism, that is, the fastening of the data unit's stator to the stator of the rotatable supporting structure (the supporting ring rigidly attached to the foundation), and the data unit's rotor to the rotor of the rotatable supporting structure, that is, to the rotatable platform, is in this case unsuitable. It was necessary to find a structural design for the data unit's installation that would eliminate the influence of the above-mentioned radial clearance on accuracy of transmission of the angle-of-turn from the platform to the rotor of the data unit. Accordingly, in order to get high accuracy, we developed a special design of data-unit mounting that eliminates the influence of the above-mentioned displacements.

The structure (Fig. 2) consists of the following basic parts: central stub 1 (column or pillar), outer mounting 2, upper housing 3, and a system of lateral adjustment of the central stub and the upper housing. The central stub is rigidly fastened to the foundation at the center

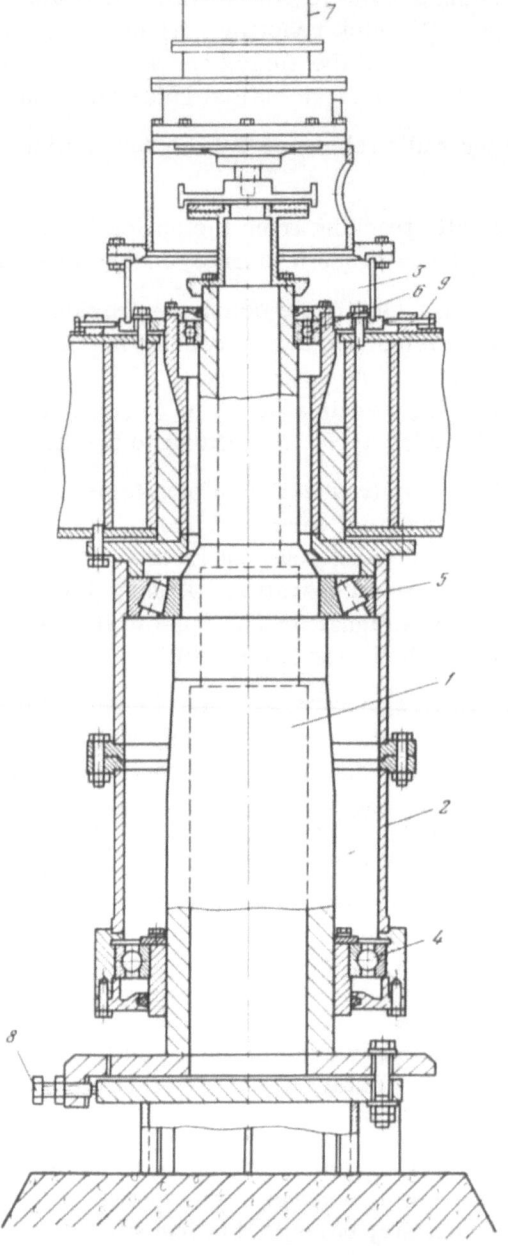

Fig. 2. Structure of the central stub.

of the rotatable structure. The upper end of the central stub serves for connection with the immobile part (the stator) of the data unit by means of a firm coupling that has no free play. The outer mounting is rigidly fastened to the rotatable platform on the underside of its floor, which is in the form of a double diaphragm. The articulation of the central stub with the outer mounting is accomplished by means of two radial bearings: of balls (No. 4 in the lower part) and of conical rollers (No. 5 in the upper part), spaced 60 cm apart in height.

The upper housing is also rigidly fastened to the floor of the rotatable platform on its upper side and articulated with the central stub by a third radial ball bearing 6. The outer mounting 7 of the upper housing serves for fastening the mobile part (the rotor) of the unit. The centering of the immobile stub, and of the outer mounting and upper housing which are mobile, is accomplished by means of the system of lateral adjustment.

The system of lateral adjustment consists of two groups of adjusting screws. The lower group of screws (which are three) is located in the lower part 8 of the central stub and ensures the central mounting of the stub. The upper group of screws is located on the upper side of floor 9 of the rotatable platform and ensures the centering of the upper housing with respect to the central stub.

When the platform rotates, the above-described structure of data-unit mounting does not hinder the possible lateral displacements of the platform, but absorbs them by means of flexure of the central stub. The design provides for the possibility of reducing the lateral displacements of the central stub's axis and of the data unit's axis by eliminating radial clearance or play in the ball bearings by means of a mutual axial displacement of the inner and outer ball races. The presumable error in the transmission of the azimuth angle due to the data unit's mounting mechanism will not exceed 5".

At the present time, the equipment for the system of automatic guidance of the Physics Institute's RT-22 and the adjusting devices for the data units have been fabricated, an M-6000 electronic computer has been installed and put into operation with the radio telescope, and preparatory work in the assembly of the new equipment is being done.

LITERATURE CITED

1. P. D. Kalachev and A. E. Salomonovich, Tr. FIAN, 17:13 (1961).
2. V. I. Ivanov, I. G. Moiseev, and Yu. G. Monin, Izv. Krym. Astrofiz. Observ., 38:141 (1967).
3. A. V. Kutsenko, B. A. Polos'yants, Yu. M. Polubesova, and R. L. Sorochenko, Automation of Radio-Astronomic Research with the RT-22 Radio Telescope by Means of an Electronic Computer. Report at the 8th All-Union Conference on Radio Astronomy, Pushchino (1975).
4. G. N. Palii, Yu. A. Fedorov, Yu. P. Ilyasov, A. P. Smirnov, and N. S. Solomin, Service of Unique Time and Frequency of the Radio-Astronomic Standards of the Physics Institute of the Academy of Sciences of the USSR. Report at the 8th All-Union Conference on Radio Astronomy, Pushchino (1975).

ANALYSIS OF THE COMPONENTS OF AERODYNAMIC MOMENTS IN PARABOLIC ANTENNAS

V. E. D'yachkov, S. L. Myslivets, and V. P. Nazarov

The problem of the decomposition, into its components, of the aerodynamic moment of a parabolic reflector with respect to the latter's apex is solved. It is established that with zero angle of attack of the reflector, which is characteristic, as a rule, of the maximum loads on the guidance mechanism of the antenna-rotation system (ARS), the principal components are the moments from the couple of the antiparallel forces and from that of the normal force. The results presented can be used for developing methods of aerodynamically balancing the moments characteristic of antenna-rotation systems in which the antenna involves a parabolic mirror.

Wind loading is one of the principal factors that affect the capacity for work of an antenna-rotating system (ARS). By means of various measures it is attempted to reduce the magnitude of wind loads on the ARS, including the aerodynamic moments with respect to its guidance axes. A reduction of load on the guidance mechanism is possible, not only through limitation of the operating maximum of wind velocity and the erection of the system in a region that has less severe climatic conditions, but also by mounting, on the antenna, special structures called aerodynamic balancers [1].

An ARS is, in the aerodynamic respect, a combination of a parabolic reflector together with a counterreflector and an antenna exciter, a girder- or truss-like supporting structure, electronic booths, counterweights, and a rotatable supporting structure, of which elements the principal moment-forming one is the reflector. Therefore, in order to develop the most effective design of aerodynamic balancing, it is important to solve the problem of decomposing wind moment from the reflector with respect to its apex into components (Fig. 1): (1) from the tangential force $Q_{x_0} t$; (2) from the normal force $Q_{z_0} m$; and (3) from the couple of antiparallel forces $2Q'_{z_0} n$. The solution of this problem will make it possible to determine the principal moment-forming factors (of force, and geometrical) and plan how to diminish them.

The solution of the problem is possible once we know the distribution of wind pressure on the surface of the parabolic reflector. Furthermore, knowledge of the pressure distribution helps to explain the physical pattern of what is occurring, verify the correctness of the adopted design, and exactly design the elements of the parabolic mirror's metallic structure for rigidity and strength.

Pressure curves, or diagrams, are obtained both from models by the drainage method (widely employed in applied aerodynamics [2]) and from actual ARS by means of special pressure gauges [3]. We must give preference to the testing of model mirrors in wind tunnels as more convenient and at the same time more accurate, in view of the considerable disadvantages of testing real ARS under natural conditions. These disadvantages, in particular, are the follow-

53

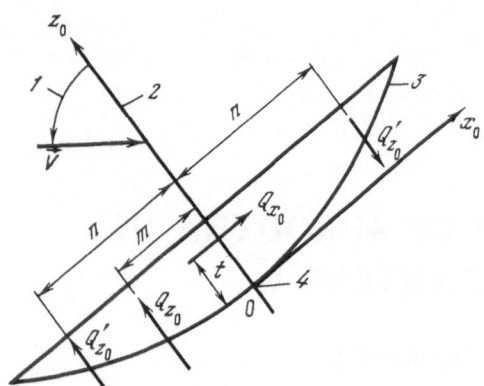

Fig. 1. Moment-forming factors (Q_{x_0}, Q_{z_0}, Q'_{z_0}, t, m, n) of the system of aerodynamic forces acting on a parabolic reflector. 1) Angle of slip of the wind; 2) focal axis of the reflector; 3) the reflector; 4) the apex of the reflector.

ing: the necessity of constantly determining the parameters of the natural wind and estimating the possible influence of errors in their determination on the result of the measurements; the large amount of labor required in conducting such tests; and other faults.

The output parameter in the experimental determination of a pressure field is the surplus pressure at a j-th point of the surface being tested, $\bar{p}_j - \bar{p}_{stat}$, where $\bar{p}_{stat}$ is the static pressure in the flow. However, in practice, the parameter more widely used is the so-called pressure ratio: $\bar{p}_j = (\bar{p}_j - \bar{p}_{stat})/q$, which is the ratio of the surplus pressure, at the point being tested, to the velocity head (kinetic-energy head) q of the undisturbed airstream before it comes close to the obstacle; $q = \rho V^2/2$, where ρ is the density of the air.

The calculation of the aerodynamic moment's components with respect to the reflector's apex is conducted under conditions of an absence of sufficiently complete and accurate data on the pressure distribution over the reflector's surface. The data that have been published can be characterized as follows:

1. The data on pressure fields that are available in the Soviet Union, discussed in reports of the LPI* [2], were obtained by testing comparatively small models in wind tunnels. These data cannot provide a complete and accurate description of the aerodynamic loads on a reflector for the following reasons: (a) the tests were carried out with a limited number of different orientations of the ARS model with respect to the direction of the airstream, and so these orientations may not have included those which could be most dangerous from the point of view of wind loading (in particular, there is an absence of data on the pressure fields of a reflector at small angles of attack); (b) the results of the tests have a considerable data spread (as much as 15%); (c) the models of the reflector were tested under conditions that did not include the influence of the earth's surface, nor that of the structures on which the mirror system is mounted.

2. The published data of foreign research [4, 5] include information on the pressure fields or influence of the earth's surface; however, they can only supplement the CAHI† and LPI investigations in a qualitative way, for their numerical data are not given in sufficient detail in order to use them, as a basis for computations.

3. From the above it follows that, for the solution of our problem, only the experimental data of CAHI and LPI can be used at this time, but corrected for the influence of the earth's surface. Investigations show that the variation of wind velocity with height above the ground is, as regards the aerodynamic loading of the reflector, a secondary factor for the case of small angles of attack.

---

* Leningrad Polytechnic Institute.
† Central Aerohydrodynamic Institute.

The present article presents determinations of the aerodynamic coefficients of tangential, normal, and antiparallel-couple forces that act on the parabolic reflector of the antenna system, and of the resulting components of the moment with respect to the paraboloid's apex, according to an overall schematic diagram of its loading. This overall schematic diagram is a set of approximated line-chord diagrams of pressure distribution over the surface of a reflector model in two mutually perpendicular diametric sections, one of which is in the plane of symmetry of the loads acting on the reflector (Fig. 2). It is assumed that the diagram of pressure distribution along a circumference r = const is also linear. The sign of the coefficient of difference in the pressures on the concave and convex surfaces of the reflector, $\Delta\bar{p}=\bar{p}_{conc}-\bar{p}_{conv}$, is positive in the direction of the external normal **n** to the surface of the mirror. For a closer approximation of the experimental diagrams of pressure distribution for any angle of attack of the reflector, analysis is conducted on the separate effects of the airflow on the windward and leeward halves of the mirror.

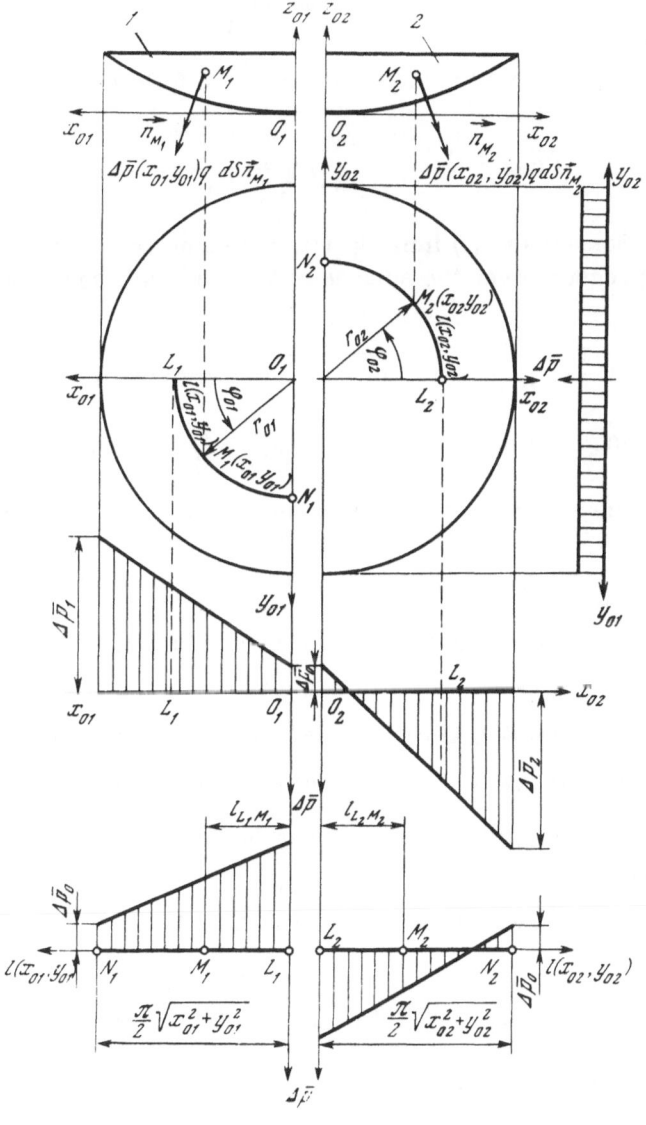

Fig. 2. Schematic diagram of pressure distribution over the surface of a parabolic reflector. 1) Windward half of the mirror; 2) leeward half.

In accordance with the symbols used in Fig. 2, the law of pressure variation along the axes X_{01} and X_{02} is of the form

$$\Delta \bar{p}(x_{0i}, 0) = \Delta \bar{p}_0 + \frac{2(\Delta p_i - \Delta p_0)}{D} x_{0i} \qquad (i = 1, 2). \tag{1}$$

Then the pressure at a point L_i will be expressed in terms of a flowing point M_i situated on a circumference of radius $r = (x_{0i}^2 + y_{0i}^2)^{1/2}$ in the following way:

$$\Delta \bar{p}_{L_i} = \Delta \bar{p}_0 + \frac{2(\Delta p_i - \Delta p_0)}{D} \sqrt{x_{0i}^2 + y_{0i}^2}. \tag{2}$$

On the basis of the assumption that, along the circumference $L_i M_i N_i$, the pressure varies also by a linear law, and taking into consideration that the length of the arc is

$$l(x_{0i}, y_{0i}) = \sqrt{x_{0i}^2 + y_{0i}^2} \arcsin \frac{y_{0i}}{\sqrt{x_{0i}^2 + y_{0i}^2}}, \tag{3}$$

the pressure at point M_i (x_{0i}, y_{0i}) will be determined by the following relationship:

$$\Delta \bar{p}_{M_i} = \Delta \bar{p}_{L_i} - \frac{\Delta p_{L_i} - \Delta p_0}{(\pi/2) \sqrt{x_{0i}^2 + y_{0i}^2}} l(x_{0i}, y_{0i}). \tag{4}$$

After substituting Eqs. (2) and (3) into Eq. (4), and some transformations, we derive the law of the variation of pressure over the surface of the windward and leeward halves of the mirror:

$$\Delta \bar{p}(x_{0i}, y_{0i}) = \Delta \bar{p}_0 + \frac{2(\Delta p_i - \Delta p_0)}{D} \sqrt{x_{0i}^2 + y_{0i}^2} - \frac{4(\Delta p_i - \Delta p_0)}{\pi D} \sqrt{x_{0i}^2 + y_{0i}^2} \arcsin \frac{y}{\sqrt{x_{0i}^2 + y_{0i}^2}} \tag{5}$$

$$(0 \leqslant x_{0i} \leqslant D/2, \quad 0 \leqslant y_{0i} \leqslant D/2, \quad x_{0i}^2 + y_{0i}^2 \leqslant D^2/4).$$

An elementary force normal to an area element dS (x_{0i}, y_{0i}) of the mirror's surface is

$$dQ_n(x_{0i}, y_{0i}) = \Delta \bar{p}(x_{0i}, y_{0i}) q dS(x_{0i}, y_{0i}) \mathbf{n}.$$

The tangential and normal components of aerodynamic force of the element of area dS are

$$dQ_{x_{0i}}(x_{0i}, y_{0i}) = \Delta \bar{p}(x_{0i}, y_{0i}) q dS(x_{0i}, y_{0i}) \cos \alpha(x_{0i}, y_{0i}),$$

$$dQ_{z_{0i}}(x_{0i}, y_{0i}) = \Delta \bar{p}(x_{0i}, y_{0i}) q dS(x_{0i}, y_{0i}) \cos \gamma(x_{0i}, y_{0i}), \tag{6}$$

where $\cos \alpha(x_{0i}, y_{0i})$ and $\cos \gamma(x_{0i}, y_{0i})$ are the direction cosines of the external normal $\mathbf{n}$ to the surface of the mirror.

The aerodynamic moment with respect to the apex of the reflector, from components (6), is

$$dM_{y_{0i}}(x_{0i}, y_{0i}) = dQ_{x_{0i}}(x_{0i}, y_{0i}) z_{0i} - dQ_{z_{0i}}(x_{0i}, y_{0i}) x_{0i}. \tag{7}$$

For the parabolic reflector as a whole,

$$Q_{x_0} = -Q_{x_{01}} + Q_{x_{02}} = q \sum_{i=1}^{2} (-1)^i \iint_{S_i} \Delta \bar{p}(x_{0i}, y_{0i}) \cos \alpha(x_{0i}, y_{0i}) dS(x_{0i}, y_{0i}),$$

$$Q_{z_0} = Q_{z_{01}} + Q_{z_{02}} = q \sum_{i=1}^{2} \int \int_{S_i'} \Delta \bar{p}\,(x_{0i}, y_{0i}) \cos \gamma\,(x_{0i}, y_{0i})\, dS\,(x_{0i}, y_{0i}), \qquad (8)$$

$$M_{y_0} = -M_{y_{01}} + M_{y_{02}} = q \sum_{i=1}^{2} (-1)^i \int \int_{S_{ij}'} \Delta \bar{p}\,(x_{0i}, y_{0i})[z_{0i} \cos \alpha\,(x_{0i}, y_{0i}) - x_{0i} \cos \gamma\,(x_{0i}, y_{0i})]\, dS\,(x_{0i}, y_{0i}),$$

where S_i' are the projection of the surface on the windward side (i = 1) and the projection of the surface on the lee side (i = 2) of the reflector onto the plane x_0, y_0.

Equations (8) are surface integrals of the first type, defined by the known pressure distribution $\Delta \bar{p}\,(x_0, y_0)$ and taken over the area S of the reflecting mirror surface. In them it is necessary to develop the values of area elements $dS(x_{0i}, y_{0i})$ and of the direction cosines $\cos \alpha\,(x_{0i}, y_{0i})$ and $\cos \gamma\,(x_{0i}, y_{0i})$ normal to these area elements.

As is well known, the normal to a surface, F(x, y, z) = 0 at a point M(x, y, z) can be represented in a canonical form:

$$\frac{X - x}{\partial F / \partial x} = \frac{Y - y}{\partial F / \partial y} = \frac{Z - z}{\partial F / \partial z}.$$

As applied to a paraboloid of revolution $x_0^2 + y_0^2 - 2pz_0 = 0$:

$$\frac{X - x_0}{x_0} = \frac{Y - y_0}{y_0} = \frac{Z - z_0}{-p}. \qquad (9)$$

Equations (9) express the collinearity of vector $MM'\{X - x_0,\, Y - y_0,\, Z - z_0\}$ with direction vector $\mathbf{a}\,\{x_0,\, y_0,\, -p\}$ (Fig. 3). From this it follows that the direction cosines of vector $\mathbf{a}$, collinear with the normal at point M(x_0, y_0, z_0) of the paraboloid of revolution, are equal to the direction cosine of the vector of the normal:

$$\cos \alpha = \frac{x_0}{\sqrt{x_0^2 + y_0^2 + p^2}}, \qquad \cos \gamma = \frac{-p}{\sqrt{x_0^2 + y_0^2 + p^2}}, \qquad (10)$$

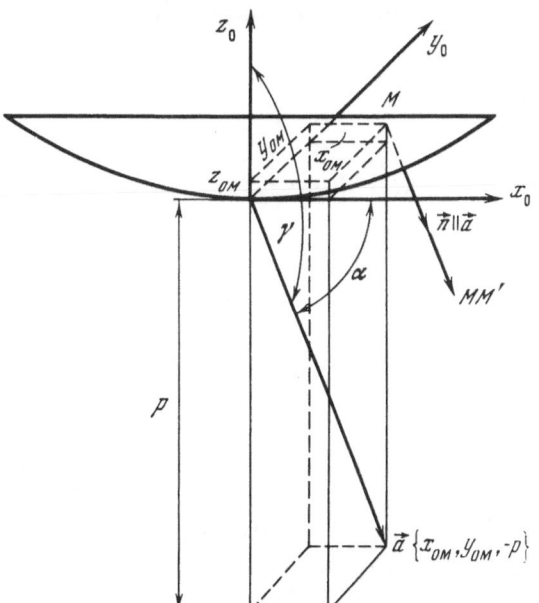

Fig. 3. Determination of direction cosines of a normal to the surface of a mirror.

where α and γ are the angles formed by the positive direction of the axes $0x_0$ and $0z_0$ with the vector $\mathbf{a}\{x_0,\, y_0,\, -p\}$; $|\mathbf{a}| = (x_0^2 + y_0^2 + p^2)^{1/2}$ is the modulus, or absolute value, of $\mathbf{a}$.

Taking account of Eqs. (5) and (10) and also keeping in mind that

$$dS = \sqrt{1 + \left(\frac{\partial z_0}{\partial x_0}\right)^2 + \left(\frac{\partial z_0}{\partial y_0}\right)^2}\, dx_0 dy_0 = \frac{\sqrt{x_0^2 + y_0^2 + p^2}}{p}\, dx_0 dy_0,$$

we rewrite Eqs. (8) in the form

$$Q_{x_0} = \frac{2q}{p}\sum_{i=1}^{2}(-1)^i \int_0^{D/2}\int_0^{\sqrt{D^2/4-x_{0i}^2}} f(\Delta\bar{p}_i, x_{0i}, y_{0i})\, x_{0i} dx_{0i} dy_{0i},$$

$$Q_{z_0} = -2q\sum_{i=1}^{2}\int_0^{D/2}\int_0^{\sqrt{D^2/4-x_{0i}^2}} f(\Delta\bar{p}_i, x_{0i}, y_{0i})\, dx_{0i} dy_{0i}, \tag{11}$$

$$M_{y_0} = \frac{2q}{p}\sum_{i=1}^{2}(-1)^i \int_0^{D/2}\int_0^{\sqrt{D^2/4-x_{0i}^2}} f(\Delta\bar{p}_i, x_{0i}, y_{0i})\left(p + \frac{x_{0i}^2 + y_{0i}^2}{2p}\right)x_{0i} dx_{0i} dy_{0i},$$

where

$$f(\Delta\bar{p}_i, x_{0i}, y_{0i}) = \Delta\bar{p}_0 + \frac{2(\Delta\bar{p}_i - \Delta p_0)}{D}\sqrt{x_{0i}^2 + y_{0i}^2} - \frac{4(\Delta p_i - \Delta p_0)}{\pi D}\sqrt{x_{0i}^2 + y_{0i}^2}\arcsin\frac{y_{0i}}{\sqrt{x_{0i}^2 + y_{0i}^2}}.$$

After replacement in Eqs. (11) of the integration variables x_{0i} and y_{0i} by cylindrical coordinates r_i and φ_i, interrelated by the correlations

$$x_{0i} = r_i\cos\varphi_i, \qquad y_{0i} = r_i\sin\varphi_i,$$

$$dx_{0i} dy_{0i} = J\, dr_i d\varphi_i = \begin{Vmatrix} \dfrac{\partial x_{0i}}{\partial r_i} & \dfrac{\partial x_{0i}}{\partial \varphi_i} \\[2mm] \dfrac{\partial y_{0i}}{\partial r_i} & \dfrac{\partial y_{0i}}{\partial \varphi_i} \end{Vmatrix} dr_i\, d\varphi_i$$

($J = r_i$ is the Jacobian substitution), and after integrating, we obtain

$$Q_{x_0} = \frac{2q}{p}\sum_{i=1}^{2}(-1)^i \int_0^{\pi/2}\int_0^{D/2} f(\Delta\bar{p}_i, r_i, \varphi_i)\, r_i^2\cos\varphi_i dr_i\, d\varphi_i = \frac{qD^3}{8\pi p}(\Delta\bar{p}_2 - \Delta\bar{p}_1),$$

$$Q_{z_0} = -2q\sum_{i=1}^{2}\int_0^{\pi/2}\int_0^{D/2} f(\Delta\bar{p}_i, r_i, \varphi_i)\, r_i dr_i\, d\varphi_i = -\frac{\pi qD^2}{24}(4\Delta\bar{p}_0 + \Delta\bar{p}_1 + \Delta\bar{p}_2), \tag{12}$$

$$M_{y_0} = \frac{2q}{p}\sum_{i=1}^{2}(-1)^i \int_0^{\pi/2}\int_0^{D/2} f(\Delta\bar{p}_i, r_i, \varphi_i)\left(p + \frac{r_i^2}{2p}\right)r_i^2\cos\varphi_i dr_i\, d\varphi_i = \frac{qD^3}{96\pi p^2}(D^2 + 12p^2)(\Delta\bar{p}_2 - \Delta\bar{p}_1),$$

where

$$f(\Delta\bar{p}_i, r_i, \varphi_i) = \Delta\bar{p}_0 + \frac{2(\Delta\bar{p}_i - \Delta\bar{p}_0)}{D}r_i - \frac{4(\Delta\bar{p}_i - \Delta\bar{p}_0)}{\pi D}r_i\varphi_i.$$

Let us proceed now to the problem of determining the magnitudes of the antiparallel forces in the direction of the axis $0z_0$ and the moment created by them. For visualizability, the pressure $\Delta\bar{p}$, which is the cause of this moment, is marked with wavy lines in Fig. 4.

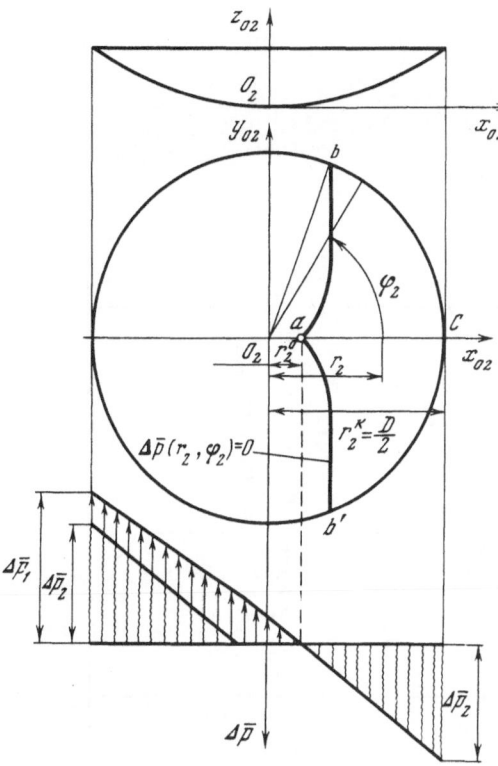

Fig. 4. Determination of the moment of an antiparallel force couple.

As follows from [5], the law of pressure variation over the surface of the mirror's leeward half, in cylindrical coordinates, is as follows:

$$\Delta \bar{p}(r_2, \varphi_2) = \Delta \bar{p}_0 + \frac{2(\Delta \bar{p}_2 - \Delta \bar{p}_0)}{D} r_2 - \frac{4(\Delta \bar{p}_2 - \Delta \bar{p}_0)}{\pi D} r_2 \varphi_2. \tag{13}$$

For deriving the equation of the curve ab, which divides the region of positive pressures from that of negative pressures, it is sufficient to set $\Delta \bar{p}(r_2, \varphi_2) = 0$, with which $\varphi_2 = \pi/2 + d/r_2$, where $d = \Delta \bar{p}_0 \cdot \pi D/4(\Delta \bar{p}_2 - \Delta \bar{p}_0)$. When $\varphi_2 = 0$, $r_2^0 = -2d/\pi = kD$, where $k = -\Delta \bar{p}_0/2(\Delta \bar{p}_2 - \Delta \bar{p}_0)$.

One of the antiparallel forces that form the force couple, in accordance with the second equation of (8), after we change the variables to cylindrical coordinates and integrate, takes the following form:

$$Q'_{z_0} = -2q \iint f(\Delta \bar{p}_2, r_2, \varphi_2) r_2 dr_2 d\varphi_2 =$$

$$= -2q \int_0^{\pi/2 + d/r_2} \int_{kD}^{D/2} \left[\Delta \bar{p}_0 + \frac{2(\Delta \bar{p}_2 - \Delta p_0)}{D} r_2 - \frac{4(\Delta p_2 - \Delta p_0)}{\pi D} r_2 \varphi_2 \right] r_2 dr_2 d\varphi_2 = -\frac{\pi q D^2}{24} \frac{\Delta \bar{r}_2^3}{(\Delta \bar{p}_2 - \Delta \bar{p}_0)^2}. \tag{14}$$

In order to determine the moment M'_{y_0} produced by the antiparallel force couple Q'_{z_0}, it is necessary to integrate the pressure field over the surface abcb' of the lee half of the mirror and double the result of the integration:

$$M'_{y_0} = 2M'_{y_{02}} = 2q \cdot 2 \iint \Delta \bar{p}(x_{02}, y_{02})[-x_{02} \cos \gamma (x_{02}, y_{02})] dS(x_{02}, y_{02}) =$$

$$= 4q \int_0^{\pi/2 + d/r_2} \int_{kD}^{D/2} \left[\Delta \bar{p}_0 + \frac{2(\Delta \bar{p}_2 - \Delta \bar{p}_0)}{D} r_2 - \frac{4(\Delta p_2 - \Delta p_0)}{\pi D} r_2 \varphi_2 \right] r_2^2 \cos \varphi_2 dr_2 d\varphi_2 = \frac{q D^3}{48} \Delta \bar{p}_0 f(\varkappa), \tag{15}$$

TABLE 1

Moment-forming factors			Results of computation of the aerodynamic coefficients of a reflector ($\bar{h} = 0.17$) and of its supporting structure according to the familiar pattern of pressure distribution with the focal axis directed to the zenith $\Delta \bar{p}_0 = -0.17$; $\Delta \bar{p}_1 = -1.01$; $\Delta \bar{p}_2 = 0.61$		Experimental data from a model in an airstream ($\bar{h} = 0.17$) with its supporting structure [6]
force factors	geometrical, referred to the characteristic linear dimension	The components of the coefficient of moment with respect to the apex of a reflector with relative depth $\bar{h}$ ($\bar{h} = 1/8\bar{p}$)			
The parabolic reflector's coefficient of tangential drag force C_{x_0} ss $= \frac{4h}{\pi^2}(\Delta p_2 - \Delta \bar{p}_1)$	$t = \frac{2}{3}\bar{h} = 0.113$	$(C_{m y_0})_{C_{x_0}} = \frac{8\bar{h}^2}{3\pi^2}(\Delta p_2 - \Delta p_1)$	$C_{x_0} = 0.112$	$(C_{m y_0})_{C_{x_0}} = 0.0127$ (12% from $C_{m y_0}$)	(C_{x_0})e $= 0.243$
The supporting structure's coefficient of tangential drag force (C_{x_0}) ss $= C_{x_0}^e - C_{x_0}$	$\bar{i}' = 0.085$	$(C_{m y_0})$ ss $= C_{m y_0}^e - \frac{16\bar{h}^2 + 3}{6\pi^2}(\Delta p_2 - \Delta \bar{p}_1)$	(C_{x_0}) ss $= 0.131$	$(C_{m y_0})$ ss $= 0.0112$ (10.5% from $C_{m y_0}$)	—
The reflector's coefficient of normal force $C_{z_0} = -\frac{1}{6}(4\Delta \bar{p}_0 + \Delta \bar{p}_1 + \Delta \bar{p}_2)$	$\bar{m} = 0.207$	$(C_{m y_0})_{C_{z_0}+C'} = \frac{1}{2\pi^2}(\Delta p_2 - \Delta \bar{p}_1)$ $(C_{m y_0})_{C_{z_0}} = (C_{m y_0})_{C_{z_0}+C'} - (C_{m y_0})'_{C_{z_0}}$	$C_{z_0} = 0.180$	$(C_{m y_0})_{C_{z_0}+C'} = 0.0821$ $(C_{m y_0})_{C_{z_0}} = 0.0373$ (35.2% from $C_{m y_0}$)	(C_{z_0})e $= 0.173$
The coefficient of anti-parallel force $C'_{z_0} = \pm \frac{1}{6}\frac{\Delta \bar{p}_2^3}{(\Delta \bar{p}_2 - \Delta p_1)^2}$	$\bar{n} = 0.361$	$(C_{m y_0})_{C_{z_0}+C'} = (1/2\pi^2)(\Delta \bar{p}_2 - \Delta \bar{p}_1)$ $(C_{m y_0})_{C'_{z_0}} = \frac{\Delta \bar{p}_0}{12\pi} f(\varkappa)$, where $f(\varkappa) = \varkappa^8\left(\frac{6 + 6\sin \varkappa}{\varkappa^4} + \frac{2\cos \varkappa}{\varkappa^3}\right.$ $\left. - \frac{\sin \varkappa}{\varkappa^2} - \frac{\cos \varkappa}{\varkappa} - 1.775 - \varkappa + \frac{\varkappa^3}{18} - \frac{\varkappa^5}{600} + \cdots \right)$; $\varkappa = \frac{\Delta \bar{p}_0}{\Delta \bar{p}_2 - \Delta \bar{p}_0}\frac{\pi}{2}$	$C'_{z_0} = \pm 0.062$	$(C_{m y_0})_{C_{z_0}+C'} = 0.0821$ $(C_{m y_0})_{C'_{z_0}} = 0.0448$ (42.3% from $C_{m y_0}$)	—
As a whole	—	$C_{m y_0} = \frac{16\bar{h}^3 + 3}{6\pi^2}(\Delta \bar{p}_2 - \Delta \bar{p}_1) + (C_{m y_0})$ ss	$C_{m y_0} = 0.106$		$(C_{m y_0})$e $= 0.106$

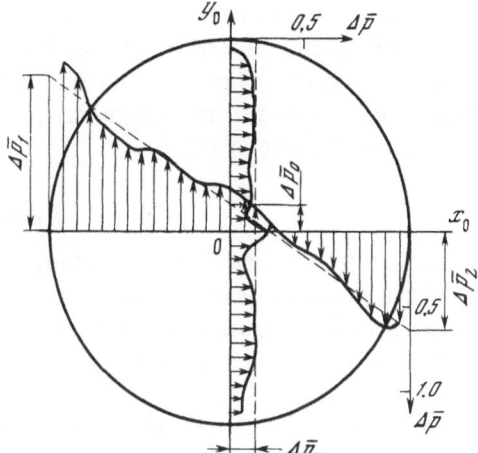

Fig. 5. Experimental and linearly approximated diagrams of pressure distribution over the surface of a parabolic reflector with its focal axis directed to the zenith.

where

$$f(\varkappa) = \varkappa^3\left[\frac{6(1+\sin\varkappa)}{\varkappa^4} + \frac{2\cos\varkappa}{\varkappa^3} - \frac{\sin\varkappa}{\varkappa^2} - \frac{\cos\varkappa}{\varkappa} - 1.775 - \varkappa + \frac{\varkappa^3}{18} - \frac{\varkappa^5}{600} + \cdots\right], \qquad \varkappa = \frac{\Delta\bar{p}_0}{\Delta\bar{p}_2 - \Delta\bar{p}_0}\frac{\pi}{2}.$$

Equations (12), (14), and (15) can be rewritten in a dimensionless form, as aerodynamic coefficients of resistance (see Table 1), relating the values of the forces to the quantity qS, and the values of the moments to qSD, where $S = \pi D^2/4$ is the characteristic area and D is the parabolic mirror's diameter, adopted as its characteristic linear dimension.

As an example, Table 1 presents results of computation of the aerodynamic drag-force coefficients of a parabolic reflector having a relative depth $\bar{h} = 0.17$ on the basis of the pressure-distribution diagram in Fig. 5, and experimental data from a model in an airstream, the model having the same relative depth and girder- or truss-type supporting structure [6], and with zero angle of attack. This orientation of a mirror with respect to the direction of the air flow (in particular, when its focal axis is directed to the zenith) is characterized, as a rule, by the maximum aerodynamic moment with respect to the reflector's apex. A comparison of the experimental and computed data allows us with sufficient accuracy (for practical purposes) to determine the aerodynamic coefficients of resistance or drag of a reflector's supporting structure with the above-stated orientation of the reflector.

CONCLUSIONS

1. The equations presented in the table make it possible to determine the aerodynamic coefficients of resistance of a parabolic reflector whatever its angle of attack, if the pressure distribution over the reflector's surface is known.

2. For an orientation of the antenna at zero angle of attack, which is characterized, as a rule, by the maximum aerodynamic-force moment with respect to the apex of the reflector, the principal moment-forming factors are the antiparallel-force couple Q'_{z_0} with its arm n, and the normal force Q_{z_0} with its arm m contributing 42 and 35% of the moment respectively.

In the development of methods for aerodynamically balancing the moments of an antenna-rotating system that has a parabolic reflector, it is important to pay special attention to the reduction of the moments formed precisely by these two factors. This may be achieved by an equalization of pressure on the surface of the reflector. The greatest effectiveness in this approach, in our opinion, is obtained by an annular flap with a gap at the back surface of the mirror. At this time research is being conducted on the determination of the optimum dimensions of such a flap. According to preliminary data, the employment of such a system of aerodynamic balancing reduces the maximum aerodynamic moments, with respect to the manipulating axes of the aiming and guidance system, by 40%, with practically unchanged values of the aerodynamic forces [1].

LITERATURE CITED

1. V. E. D'yakov and S. L. Myslivets, "Investigation of methods of aerodynamic balancing of the moment characteristics of antenna-rotation systems with parabolic mirrors," Report of the Leingrad Polytechnic Institute (1971).
2. "Aerodynamic characteristics of radio telescopes," Report of the Leningrad Polytechnic Institute (1962).
3. "More precise determination of actual wind loads on buildings; investigation of wind pressure on buildings and other structures," Report of the Central Scientific-Research Institute of Construction Structures of the USSR (1963).
4. E. Cohen, J. Vellozi, and S. S. Suh, "Calculation of wind forces and pressures on antennas," Ann. N.Y. Acad. Sci., Vol. 116 (1964).
5. T. A. Wyatt, "The aerodynamics of shallow paraboloid antennas," Ann. N.Y. Acad. Sci., Vol. 116 (1964).
6. "Aerodynamic investigations of model antennas," Report of the Leningrad Polytechnic Institute (1965).

A METHOD FOR DETERMINING THE COST OF FABRICATION OF A FULLY ROTATABLE AND INCLINABLE RADIO TELESCOPE

P. D. Kalachev, A. V. Samotsetov, and É. A. Tret'yakov

Problems of the economics of fully rotatable and inclinable radio telescopes are examined. The dependences of the cost of the fully rotatable and inclinable parabolic antennas of radio telescopes on their weight, on the height of the antenna structure, and on the technology of the construction are analyzed. The technique of cost-estimation models of radio telescopes, with allowance for losses in designing and planning, fabrication, assembly, adjustment, and operation, is examined.

INTRODUCTION

Further development of observational radio astronomy requires the building of radio telescopes possessing ever higher values of two parameters: resolution and sensitivity. At any given wavelength, an increase in these parameters depends on an increase in the size of the radio telescope's antenna.

The most widely used radio telescopes are those that have a fully rotatable and inclinable parabolic antenna. The main limitation for increasing their size is the rapid growth of the cost of building them, including difficulties of their design and of their fabrication. The difficulties of their design are due to the need of solving problems such as finding an efficient way to reduce elastic deformations due to their own weight and due to wind pressure, and also to reduce thermoelastic deformations due to uneven heating from solar radiation. The difficulties of fabrication are mainly difficulties of adjusting the reflecting surface of the parabolic antenna.

The utilization of a parabolic mirror as the antenna of a radio telescope represents, to a certain extent, the tradition of optical astronomy, in which the most widespread type of telescope is the reflector. Thus the parabolic mirror is the classical type of antenna for radio telescopes that concentrate electromagnetic energy at a focus. The development and building of other types of antennas for radio telescopes, such as, for example, the cross-shaped antenna (the Mills cross), which work by the principle of sequential synthesis [1, 2], the variable-profile antenna system (of S. É. Khaikin and N. L. Kaidanovskii [3]), and others, are results of the search for ways to build antennas for large radio telescopes more economically.

The economy of radio telescopes, in the narrow sense of the word, concerns the cost of building the radio telescope proper and of equipping it with the necessary receiving and other electronic apparatus.

The economy of radio telescopes, in the broad sense of the word, includes not only the financial and material expenditures on planning with theoretical and experimental studies of various structural design, the cost of radio receiving apparatus and of other electronic equipment, including a computer for the automation of guidance and the recording and processing of observational results, fabrication, assembly, and adjusting, but also the cost of auxiliary and service facilities and buildings (access approaches, electric power supply, water supply, drainage, etc.), and also operating costs.

However, we shall limit ourselves at first to an estimation of the cost of building a radio telescope antenna, not touching upon other expenses (in particular, the influence of the choice of a site for a radio telescope on the cost of its construction).

1. Estimation of the Cost of Building a Fully Rotatable
and Inclinable Parabolic Radio-Telescope Antenna

The cost of the fabrication, and of the assembly and erection, of a fully rotatable and inclinable parabolic radio-telescope antenna can be expressed as a function

$$c = \varphi\,(\lambda,\ G,\ h,\ T_{tech}), \tag{1}$$

where λ is the wavelength at which the radio telescope must operate, G is the weight of the rotating part, h is the antenna's height (for the type of radio telescope under consideration, the height of the antenna is practically equal to the mirror's diameter D_m), T_{tech} is the technology of the construction, and c is the cost in rubles.

The cost of fabricating metallic structures of a given complexity is usually determined as a product

$$c_{tech} = c_1 G, \tag{2}$$

where c_1 is the cost of fabricating 1 ton, and G is the weight of the metallic structure.

Here it should be noted that Eq. (2) involves a well-known contradiction: that in order to reduce the cost of fabricating a metallic structure, one must try while designing it to reduce its weight; however, as a rule, a lightweight construction involves the use of the lightweight cross-sectional shapes of rolled stock or tubes with more complex junctions for the joining of the structural pieces and a greater number of connections, with the result that it is more complicated and so the cost of fabricating each ton of the structure is found to be higher. Accordingly, design engineers are induced to search for new basic solutions to the problems of light structures without complicating them. The structures, in particular, are those of fully rotatable and inclinable parabolic radio-telescope antennas that ensure matched deformations [4] or "homologous" deformations [5].

For a given structural design, the weight of the rotatable part of the parabolic antenna is related to its size by the correlation

$$G = k_2 D^m, \tag{3}$$

where k_2 is the proportionality factor and D is the diameter of the mirror.

According to data presented in [6], $k_2' = 28$ and m = 2.5. In this reference, the topic of discussion is only the mirror system. The weight of the rotatable supporting structure is approximately double that of the mirror system, so that $k_2 = 3k_2'$.

According to data in [7], m = 3, but with allowance for the fact that the weight of the mirror's reflecting surface (of panels or screens) varies in proportion to the square of the mirror's diameter, m can be assumed between 2 and 3.

Obviously, Eq. (3), in conformity with both [6] and [7], can be assumed acceptable for a preliminary estimate of the weight of the antenna structure of a radio telescope being planned anew according to the known weight of a prototype, but with unchanged structural design. In this case Eq. (3) takes the form

$$G_1 \approx G_0 \, (D_1/D_0)^{2.5}, \tag{4}$$

where G_1 is the weight of the newly planned radio telescope; D_1 is the diameter of its parabolic mirror; and G_0 and D_0 are, respectively, the weight and the diameter of the prototype's parabolic mirror.

The dependence of the assembly and erection cost c_{erec} on the height h of the antenna structure can be written

$$c_{erec} = k_3 h^n, \tag{5}$$

where k_3 is the proportionality factor, and n > 1.

In conformity with Eq. (5), the quantity h involves a corresponding weight; that is, generally speaking, with increase in the height of a radio telescope, its weight increases also. Thus, with n > 1, the cost of the erection labor increases faster than the increase of the weight [compare with Eq. (2)]. This is explained by the rapid rise in the cost of the lifting and carrying elevators, cranes, etc., that are needed for tall construction projects.

The dependence of the antenna structure's cost c on the wavelength λ to be used can be expressed by the relationship

$$c_\lambda = k_4 \lambda^{-\omega}, \tag{6}$$

where k_4 is the proportionality factor, dependent on the conditions of the mirror's fabrication, assembly and erection, and adjustment; the exponent ω depends on the ratio D/λ, on the absolute amounts of the tolerances in the fabrication of the mirror and of the drive mechanisms, and on the mirror's adjustment technology. In any case $\omega > 1$.

The dependence of the antenna's cost on technology of construction T_{tech} can be represented in a general form as the function

$$c_{tech} = f(G, \ J, \ M, \ R, \ \alpha, ..., \beta), \tag{7}$$

where G is the geometry of the supporting structure, J is the complexity of the junctions, M is the employment of mechanical fabrication, in particular, of large junctions, R is the use of fabricated (rolled) stock, and $\alpha, ..., \beta$ are other factors.

Under the control of the design engineer there are only two cost-determining factors: antenna weight [Eq. (3)] and construction technology [Eq. (7)], since the size and, consequently, the height of the antenna structure, and also the wavelength to be used, are already specified.

As has already been stated, in order to surmount the obstacles that stand in the way of increasing the antenna size of fully rotatable and inclinable radio telescopes, it is necessary to find new structural designs of radio-telescope antennas. Some methods for solving this problem are examined, for example, in [8].

2. Estimation of the Cost of a Radio Telescope as a Complex (the Construction of a Radio-Telescope Cost Model)

At the earliest stages of radio-telescope design, it is necessary, with the least possible error, to predict the future expenses for all the steps of designing, building, and operating it. This is necessary for securing financing and industrial supplies, for planning the work, and also for deciding on the advisability and economic feasibility of carrying out the radio-telescope project.

On the basis of the design specifications, which specify the values of the radio telescope's basic parameters (resolving power, sensitivity, wavelength λ, diameter of the parabolic mirror, and also the constitution and characteristic curve of the radio-receiving apparatus and equipment), it is determined what the radio telescope's structural parameters must be: the design of the mirror and of the rotatable supporting structure, their weight, the tolerances in the fabrication of metallic structural parts and mechanisms, and the tolerances in the adjustment of the mirror's reflecting surface. It is obvious that these structural parameters directly influence the cost of the radio telescope.

The cost of a radio telescope can be represented in the form of a cost model

$$EI = f(REP) \tag{8}$$

where EI are economic indicators and f is the sought function of the radio telescope's specified radio-engineering parameters (REP).

If we construct the radio-telescope cost model suggested in Eq. (8), then with its help it will be possible to predict the expenses of building and operating a radio telescope at the earliest stages of its designing, to determine the optimal variant of the specified REP, and to decide whether to carry out the plans.

Among the familiar methods of cost prediction (expert estimation, analogs, calculation), cost modeling is distinguished by its quick determination of expenses and its objectiveness and taking account of many incidental or accidental factors of industrial and organizational nature. Expense modeling in various fields of technology has shown the great accuracy of this method as compared to others, and at the present time it is the most scientific method. The mathematical modeling of costs requires the gathering of a large volume of statistical data on previously built radio telescopes and their structural units, an analysis of the interrelation of the radio-engineering parameters, the structural parameters, and the economic indicators, with mathematical formulas of these relationships, all of this requiring much effort and time.

The mathematical modeling of the cost of radio telescopes as a function of their radio-engineering and structural parameters has, in recent years, attracted more and more attention of domestic and foreign radiomechanical engineers. Sokolov [7] examined the problems of determining the cost of fully rotatable and inclinable radiotelescopes as a function of the wavelength to be used and presents theoretically based cost models in a general form. In publications of foreign authors, for example [9], separate cost models are given for different parts of a radio telescope, the following being of interest also to our own specialists:

$$
\begin{aligned}
c_1 &= 24.2 + 1.36D + 0.132D^2 + 0.0061D^3, \\
c_2 &= -32.9 + 72.3/T + 105.4/T^2, \\
c_3 &= 6.2 + 0.99D + 0.002D^2,
\end{aligned}
\tag{9}
$$

where c_1 is the cost of the parabolic antenna (in thousands of dollars), D is the diameter of the reflector (in feet), c_2 is the cost of the parametric amplifier (in thousands of dollars), T is the noise temperature of the receiver (in degrees Kelvin), and c_3 is the cost of the antenna exciter (in thousands of dollars).

In [5], the effect of the reflecting surface's accuracy and of the reflector's diameter on the cost and weight of the paraboloid is examined. It is shown that, with decrease in the distortion allowed in the reflecting surface and with increase in the reflector's diameter, there is a pronounced increase in the weight of the reflector's framework and, consequently, also in the cost of fabrication and of assembling and erecting the radio telescope.

An analysis of the literature has shown that there is as yet no sufficiently complete cost model even for one of the types of radio telescopes. There are individual cost calculations for radio telescopes and theoretical predictions about the influence of radio-engineering parameters on the cost, which can be used for constructing the model.

2.1. Procedure for Formulating a Cost Model

By the term c o s t of a radio telescope, we understand a final expression of all the types of expenses for the design, building, and operation (for, let us say, 10-20 years) of the radio telescope generated by all the organizations involved.

A radio-telescope cost model can be written in the general form

$$c = c(\text{SP, REP}) = \sum_{i=1}^{m} \sum_{j=1}^{n} \sum_{k=1}^{l} c_{i,j,k} (\text{SP, REP}) \tag{10}$$

where $c_{i,j,k}(\text{SP, REP})$ is an elementary cost model that expresses the expenses of the k-th enterprise in the j-th stage of the overall job on the i-th component part of the radio telescope as a function of the structural parameters (SP) and the radio-engineering parameters (REP).

Such a representation corresponds to current concepts about the cost of technical systems [10] and makes it possible to most completely represent all the necessary expenses for ensuring the functioning of a radio telescope during the specified term of service (10-20 years).

The creation of a cost model proceeds under the influence of two contradictory requirements: The model must adequately represent the complex interrelations between the selected parameters and the cost, have the smallest possible error in its predictions, and at the same time be sufficiently simple mathematically. Structurally, the model should not be more complex than is required for the given accuracy of its input data, since the effect of the permissible errors in the specified parameters and input data will nullify much work on any superfluous complexity of the model.

A difficult factor in the construction of models is the absence of the necessary statistical information, which leads to a simplification of the models, an increase in the errors of prediction, and an impoverishment of the results of investigation with such models. However, we must not refrain from using the models because of their imperfection, since the reality will always be more complex than even the finest models.

For the construction of a sufficiently accurate model, it is necessary first of all to investigate the type of radio telescope to be built and to distinguish its characteristic features and makeup. The greater the degree of project development, the more informative will the model be, and the more factors will pass from the category of the undetermined to that of the determined. Most accurate will be the model constructed from actual expenses for building one or several telescopes of the same type.

For the construction of a cost model, it is unsuitable to employ separately either the statistical method, or the analog method, or the calculation method. A combined method has been developed, one utilizing the merits of each one and capable of surmounting the main obstacle: the distinctive features and characteristics of each new radio telescope.

The basic propositions of the combined method of radiotelescope cost-model construction are the following:

1. Each radio telescope being planned consists of many structural parts, or units, sufficiently close analogs of which can be found in previous developments and existing radio telescopes.
2. The factual data on the cost of the stages of work collected from previous developments and existing radio telescopes serve as a statistical basis for the construction

of elemental cost models of the individual units of a new radio telescope as a function of its structural parameters.

3. Taking account of the theoretical relations, we can express each structural parameter of a radio telescope in terms of the basic radio-engineering parameters and derive elemental cost models as a function of these parameters in the form

$$c_{i,j,k} = c_{i,j,k}(\text{REP}). \tag{11}$$

4. Algebraic summation of the elemental models gives us the radio telescope's cost model.

In this way, the essence of the method consists in a conventional division of a radio telescope into its constituent structural parts or units, the construction of elemental models, and the algebraic summation of these models.

As factors in a mathematical cost model, we can use the basic radio-engineering and structural parameters that characterize the type of radio telescope under consideration. Functionally between themselves they do not depend, and they most strongly affect the cost; they are the wavelength, resolving power, noise temperature, diameter of the paraboloid mirror, etc. Not all radio-engineering parameters can be classified as basic; some of them can be set at some mean value, and others can be disregarded. Analogously, consideration is given to the basic structural parameters of the radio telescope, the ones that directly affect its cost.

In the construction of elemental and complete cost models, it is necessary to take into account the expenditures in all the stages of the work, including design, building, and operation.

An algorithm of radio-telescope cost-model construction is presented as a block diagram, in which the individual blocks represent the following operations (see Fig. 1):

1. Input data: type and structural design of the radio telescope, and the specified radio-engineering parameters. It is assumed that, within the specified radio-telescope design (general design), the radio-engineering parameters may be altered within specified limits.

2. Basic radio-engineering parameters; analysis of their interrelations and degree of influence of the cost; selection of the basic parameters; determination of the functional relationship of the basic radio-engineering and structural parameters.

3. Basic structural parameters (size of antenna, length of transmission lines, etc.); selection according to the following criteria: direct functional dependence on the basic parameters and influence on the cost.

4. Permissible limits of alteration of the basic structural parameters: determination of the region suitable for the operation of the specified radio-telescope design type. A falling outside the limits may indicate the expediency of a change to another structural design.

5. Permissible limits of alteration of the basic radio-engineering parameters. They are determined as a function of the limits of alteration of the basic structural parameters with allowance for the functional relations. The permissible limits of the basic parameters limit the range of operation of the desired cost model in the multivariate factor space.

6. Matrix of costs. In a structural-economic analysis a determination is made of the possible specific types of expenditures in all the stages of the work and for all the component parts or units of the radio-telescope. Examples of jobs and expenditures: radio-engineering development and design, fabrication, transportation, assembly and erection, tuning, adjustment (aligning), construction work, energy consumed, expenses of maintenance and repairs, salaries of operating personnel, etc.

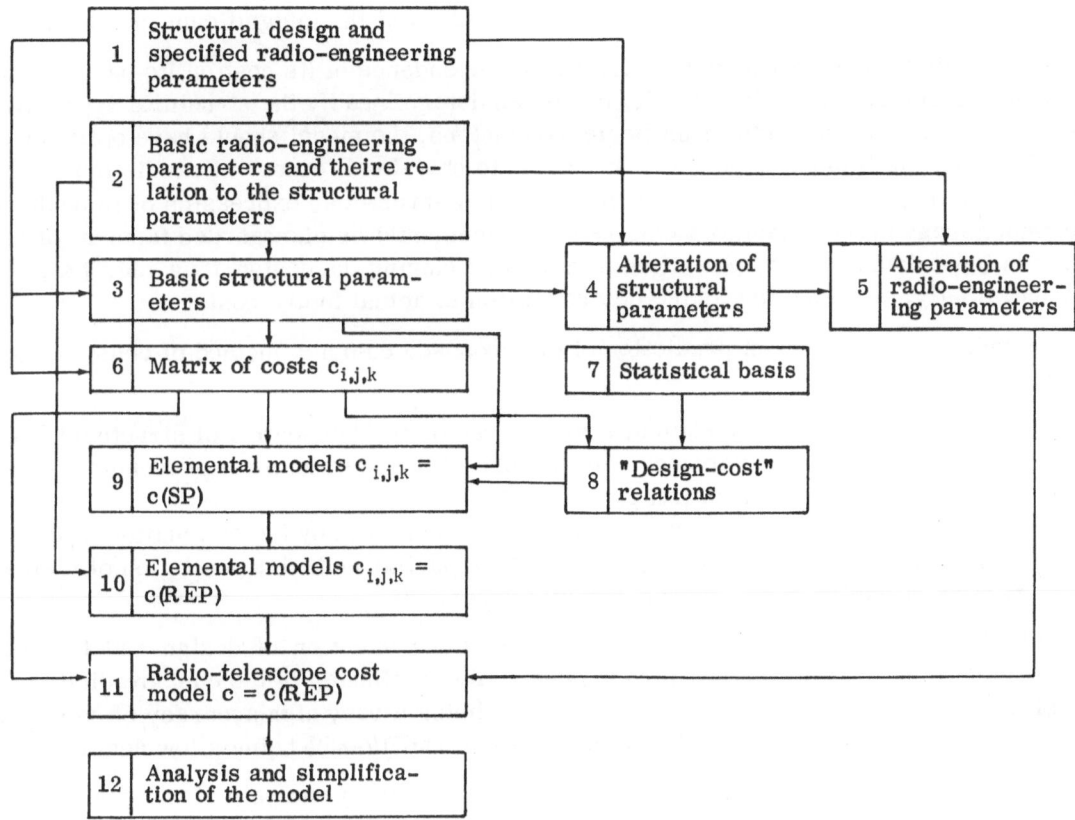

Fig. 1. Block diagram of model construction.

7. Statistical basis. A systematized and mathematically processed stock of real data, structural and economic, on the greatest possible number of radio telescopes already built. The statistical basis is a card catalog of the data on the component parts of radio telescopes, on the separate stages of the work (work estimators), on the equipment, the statistical relations found between the structural and economic characteristics of the units and manufactured articles, etc. The statistical basis is a scientific basis for the prediction of costs and the construction of a cost model.

8. Selection of the necessary statistical material (analogs, analytic dependences, procedures, etc.) for constructing elemental cost models.

9. Elemental cost models. They are constructed with a complex analysis of the projected radio telescope on the basis of the available statistical material. The radio telescope's basic structural parameters, which are directly related to the costs, are used as factors in these models.

10. Elemental cost models in which the factors are the basic radio-engineering parameters. They are derived by substituting in the previously found elemental models (paragraph 9), instead of the basic structural parameters, their expressions in terms of the basic radio-engineering parameters (paragraph 2).

11. Algebraic summation of all the elemental models, which gives the cost model of the radio telescope.

12. Simplification of the cost model by the exclusion of the insignificant members.

A radio telescope's cost model describes the dependence of its cost on its basic radio-engineering parameters, in a given family or series determined by the structural type. Within the range of possible increments in the basic parameters, the model should have optimum accuracy of prediction, it being impossible, however, to obviate all errors of prediction and construct an absolutely accurate model. Model-prediction error is an expression of imperfection and incompleteness in the model, of the influence of accidental or disregarded factors, of inaccuracy in input data and in the specified parameters (during the design they may be altered), and, finally, of inaccuracy in the determination of actual future costs.

For reducing the errors of prediction of the proposed complex method of constructing a model, there is provision for:

(a) statistical analysis of the actual expenditures on the fabrication of structural parts by several different manufacturers and the use of the average level of technical development of each branch of industry;

(b) the representation of the radio telescope's overall cost by the summation of the elemental models, which permits, as a result of partial mutual cancellation of their errors, a reduction of the overall model's error.

Analysis of the literature and practical work in the prediction of design and fabrication costs permits us to propose the following qualitative estimates of the cost-prediction accuracy of models: root-mean-square error of 3-7%, being a very high accuracy of some models; 7-10% (a high accuracy); 10-17% being the average; and 17-25% being a low accuracy.

The accuracy of model construction depends on the correct analysis of the given type of radio telescope, on the choice of the basic parameters, and on the existence of the necessary statistical basis.

2.2. Elemental Cost Models

In the process of gathering and analyzing the structural and economic information on fully rotatable and inclinable parabolic radio telescopes, one finds certain statistical relations that can be utilized in the construction of an overall cost model.

A c o u n t e r - r e f l e c t o r of the hyperbolic type, made of aluminum sheets and profiles (shapes) with stamping and riveting, has a maximum deviation from the theoretical surface

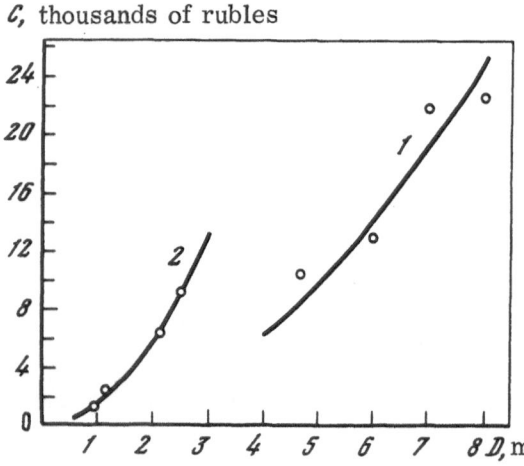

C, thousands of rubles

Fig. 2. Cost of fabrication of antenna mirrors. 1) Parabolic aluminum, $c = 0.4D^2$; 2) hyperbolic aluminum (counter-reflectors), $c = 1.5D^2$.

amounting to ±0.5 mm. The counter-reflector is fabricated with the help of special equipment and is structurally a single riveted unit that cannot be taken apart. The cost of fabrication of such round hyperbolic reflectors is well described by the statistical formula $c = 1.5D^2$, where c is the cost in thousands of rubles and D is the diameter in meters (Fig. 2).

The quadratic dependence of cost on diameter is observed also for round parabolic reflectors of 4-8 m in diameter, as is seen in Fig. 2: $c = 0.4 D^2$.

Parabolic reflectors are also fabricated from aluminum sheets, profiles (shapes), and castings with the use of stamping and riveting. The reflectors have a surface accuracy of ±1.0-1.5 mm and are fabricated with special equipment. For transportation, reflectors can be dismantled into their individual panels.

The structure of the aluminum counter-reflector and parabolic reflector is a set of radial ribs, annular diaphragms, and one edging of 1-2 mm in thickness. The quadratic dependence of cost diameter shows that the unit cost for 1 m^2 of reflecting surface is constant both for the counter-reflector and for the parabolic reflector.

The round parabolic reflectors with a diameter greater than 8-12 m have, as a rule, a fundamentally different structure: a supporting steel framework of tubular members and reflecting aluminum panels attached to them that are made to have a deviation from the theoretical surface not exceeding of ±0.5 mm. The panels are adjusted with advancement on mountings that are of adjustable length, with an accuracy of ±1 mm.

The reflecting panels are fabricated as riveted structures of aluminum sheets and profiles (shapes). The operating sheathing, of 1.5-2 mm sheets, is formed on special forming equipment and after their installation the panels are aligned. The configuration of the panels

TABLE 1

Mirror		Panels					
diameter, m	aperture, m^2	number of panels	weight, tons	weight per m^2, kg	thickness, mm	cost of fabrication, thousands of rubles	cost per m^2, rubles
16	200	136	2,74	13,7	120	30,4	152
25	490	361	5,60	11,4	160	72	147
32	800	930	12,00	15,0	100	125	156
64	3200	1236	41,00	12,8	205	503	157

TABLE 2

Diameter of reflector, m	height, m	weight, tons	cost of fabrication, thousands of rubles	cost per m^2 of aperture, rubles	structural characteristics	type of production
16	6	14	105	525	Flat radial girders and annular ones	Single
25	10	36	138	282	Central drum and three-dimens. sectors	Mass
25	10	33	130	253	Same (lighter variant)	Single
32	7	37	182	228	Flat radial girders and annular triang. ones	Mass
64	16	300	870	272	Radial and annular three-dimens. girders	Single
130	27	1700	4230	330	Central drum and three-dimensional sectors	Advance project

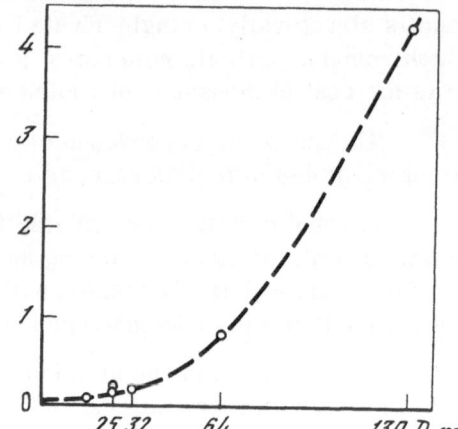

Fig. 3. Influence of the mirror's diameter D on the fabrication cost of its framework, c = $0.615D^{2.28}$ (in thousands of rubles).

is a segment whose size depends on the spacing of the framework's stiffening members. An analysis of the structural-economic data on the fabrication of the panels (Table 1) shows that their mean unit weight is 13.2 kilograms per square meter of aperture of the mirror and that the mean unit cost of fabrication of the panels is 153 rubles per square meter of aperture. These values are constant for radio telescopes of different diameters.

The weight and cost of fabrication of a parabolic reflector's framework depends on its diameter, on its structural design, and on the permitted distortion of its reflecting surface under the action of the destabilizing factors of its weight, of wind and ice loads, of nonuniform heating by the sun, and of inertial loads. According to American data, the cost of a radio telescope is proportional to $D^{2.8}$; an analysis of actual and project-estimated costs of Soviet radio telescopes gives a somewhat different dependence. The structural-economic indicators of reflector frameworks (Table 2) give the following cost model:

$$c = 0.62D^{2.28}, \tag{12}$$

where c is the cost of fabrication of the framework in thousands of rubles, and D is the diameter of the paraboloid in meters (Fig. 3).

Taking into account that the expenditures for the assembly and erection of the reflector and for the fabrication of the remaining parts of the radio telescope are approximately proportional to the square of the diameter, it may be assumed that the dependence of the overall cost for a parabolic reflector will be proportional to D^n, where n = 2.1-2.2.

The practice of radio-telescope engineering, and of these telescopes' operation, ever more pressingly poses the problem of cost prediction at the earliest stages of their planning and design. The creation of a series of cost models for the most common types of radio telescopes will enable us to solve this problem. For this purpose it is necessary in the first place to gather detailed information on the designs and costs of existing radio telescopes and then investigate the influence of individual parameters on cost. The proposed algorithm for cost-model construction can serve as a methodological basis for this task.

LITERATURE CITED

1. J. P. Wild, Usp. Fiz. Nauk, 92:706 (1967).
2. V. V. Vitkevich and P. D. Kalachev, Tr. FIAN, 28:5 (1965).
3. S. É. Khaikin, N. L. Kaidanovskii, A. N. Esepkina, and O. N. Shivris, Izv. Glav. Astron. Observ., Vol. 21, No. 5(64) (1960).
4. P. D. Kalachev, Preprint FIAN, No. 171 (1968).

5.   S. Hörner, Astron. J., 72(1):1346 (1967).

6.   B. A. Garf, Utilization of Solar Energy [in Russian], Izd. Akad. Nauk SSSR, Moscow (1957), p. 62.

7.   A. G. Sokolov, Technical-Economic Characteristics of Fully-Rotatable Radio Telescopes. Metallic Structures [in Russian], Stroiizdat, Moscow (1966), p. 140.

8.   P. D. Kalachev, Problems of Radioelectronics, Ser. General Techniques [in Russian], No. 11 (1969), p. 55.

9.   C. L. Cuccia, Microwaves, 8(6):86 (1969).

10. A. V. Glichev, Economic Efficiency of Technical Systems [in Russian], Ékonomika, Moscow (1971).

AUTOMATION OF THE PROCESSING OF
PULSAR OBSERVATIONS

B. V. Vyzhlov, V. V. Ivanova, V. A. Izvekova,
A. D. Kuz'min, Yu. P. Kuz'min,
V. M. Malofeev, Yu. M. Popov,
N. S. Solomin, T. V. Shabonova,
and Yu. P. Shitov

With the DKR-1000 and BSA meter-band radio telescopes of the Physics Institute used for the study of pulsars, digital computing technology is employed for processing the results of observations in real time. For enhancing sensitivity and time resolution, a procedure of pulse buildup with prior compensation of a pulsar's dispersion over a wide band is employed. A study of known pulsars is conducted with the help of analyzers of the "Neutron" type. To search for unknown pulsars, a system has been developed that processes observations with the on-line method, using an M-6000 electronic computer.

The experimental study of pulsars involves the reception, storage, and processing of a large number of observational data incoming per unit time. This characteristic of pulsar observations, due to the rapidly varying nature of their radio emission (a periodic sequence of pulses of short duration and varying intensity), in a majority of cases requires, for enhancing the effectiveness of their analysis, the digital recording of the signal with subsequent computer processing. Of decisive importance is the automation of observation processing when searching for and studying weak pulsars, when the individual pulses of a pulsar are much weaker than the intrinsic noises of the radio telescope, in which case it is necessary to use special methods for separating the periodic signal from the noise.

Recently at the radio-astronomical station of the Physics Institute of the Academy of Sciences, digital computing technology has been widely used in the processing of observational data. Thus, in the study of known pulsars (with a known period of repetition of their pulses), effective use has been made of the special "Neutron" analyzer with which it is possible to study "trans-noise" periodic processes and, in particular, to carry out synchronous accumulation of the pulses from pulsars in the real time scale. The accumulation of the pulses increases the sensitivity of the observations by $\sqrt{N}$ times, where N is the number of pulses accumulated. The "Neutron" analyzer has a memory capacity of 256 eight-digit elements and permits pulsar-pulse buildup either in a single channel or simultaneously in two, four, or eight channels. This makes it possible to study pulsars simultaneously at different frequencies either with different transmission bands of the radiometer, or with different polarizations. The "Neutron" is being used by us for the measurement of the following pulsar parameters: (1) the shape of the mean pulse profiles; (2) the mean energy of the radio emission and its spectrum; (3) the periods of

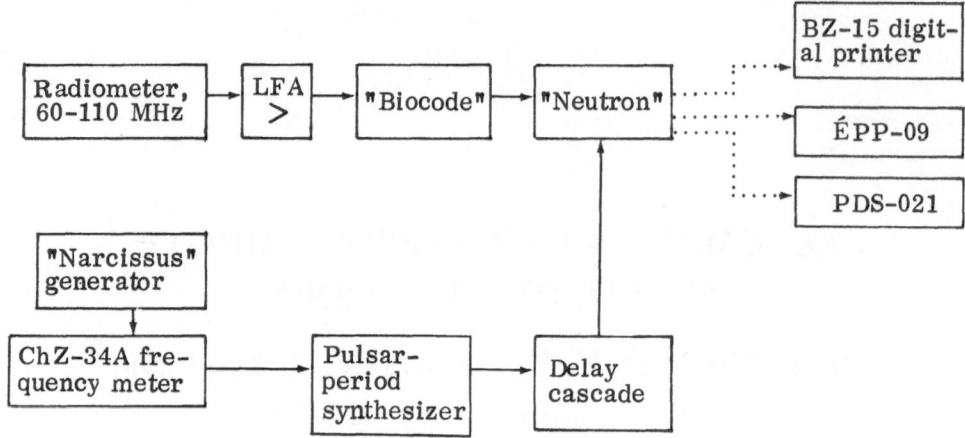

Fig. 1. Block diagram of the apparatus for observing pulsars by the method
of the synchronous accumulation of pulses.

the pulsars and the rate of change of these periods with time; (4) the polarization of the pulsars' radio emission.

A block diagram of the apparatus for observing pulsars by the method of pulse buildup in real time is presented in Fig. 1. The signal from the radiometer's output is amplified to about the 1-V level of a low-frequency amplifier (LFA) to the extent that the recording of pulses from the pulsar can be done without a constant component of radiometer noise. In this case it is easy to obtain the necessary dynamic range of the amplifier and there is no problem of a zero drift. The band of amplifiable frequencies of a low-frequency amplifier is from 0.3 to 10 kHz. Then the signal is transformed by an analog-to-digital converter of the "Biocode" type into an eight-digit binary code and enters a "Neutron" analyzer. The synchronous accumulation of pulses from pulsars takes place with the start-up of the "Neutron's" operation in the waiting regime or hold mode from a circuit that forms starting pulses with a period equal to the period of the pulsar being studied, the accuracy of the period's determination being within 5×10^{-6} sec. The observations are being conducted with the BSA and DKR-1000 meridional radio telescopes, and therefore the number of accumulated pulses is determined by the time of the pulsar's passage through the antenna's directional diagram or beamwidth, and by the period of the pulses' repetition. The number of pulses that can be accumulated for various known pulsars can range from 50 to 2000, which makes it possible to obtain a gain in sensitivity, as compared to that of recording individual pulses, of 7 to 45 times respectively.

The time resolution when studying average forms or shapes of pulses from pulsars is determined by the interval between signal readouts and can be as good as 0.1 msec in the single-channel procedure with the "Neutron."

Visual observation of pulsars during pulse buildup is accomplished with the screen of an electron-beam tube built in as an integral part of the "Neutron," and, after completion of the buildup, the shape of the pulse can be luminesced on the screen. The result of processing is stored in the "Neutron's" memory and can be led out to external digital or analog recording devices. In our observations we use a digital-printing device of the BZ-15 type, an ÉPP-09 automatic recorder, and a two-coordinate PDS-021 automatic recorder.

An example is given of the recording of the pulsar PSR 0950 + 08, in the same form as it was received on the "Neutron's" screen, in Fig. 2. An example of the recording of one of the most feeble pulsars, in the meter range, is shown in Fig. 3. The recording was made with

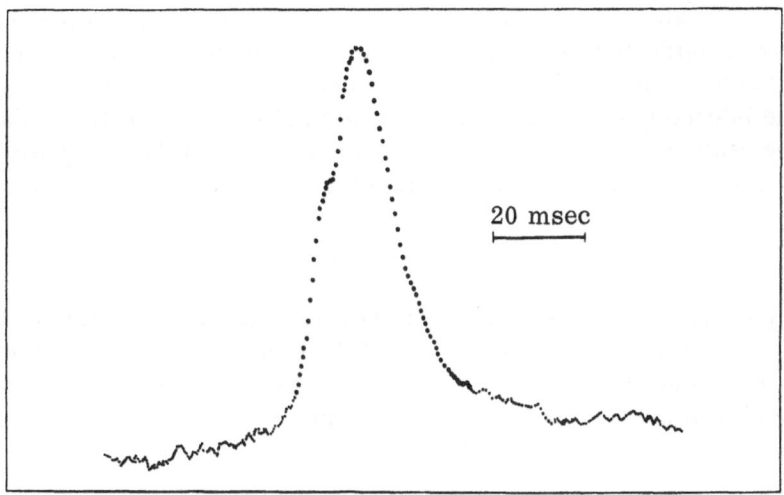

Fig. 2. Example of recording of the pulsar PSR 0950 + 08, made with the help of a "Neutron" analyzer at a frequency of 102.5 MHz. Number of pulses N = 820; interval between readouts 0.5 msec; time constant τ = 2.4 msec; reception band Δf = 25 kHz.

the help of the new BSA highly sensitive antenna [1]. Without accumulation of pulses, this pulsar is not observable.

In addition to having an increase of buildup time, the sensitivity of reception of a pulsar's radio emission can be augmented by having a wider transmission band in the radio receiving apparatus. However, in view of the pulsed nature of the radiation from pulsars, the width of the reception band is limited by dispersion in the interstellar medium. The velocity of propagation of radio waves in the interstellar plasma is less than in free space, and depends on the

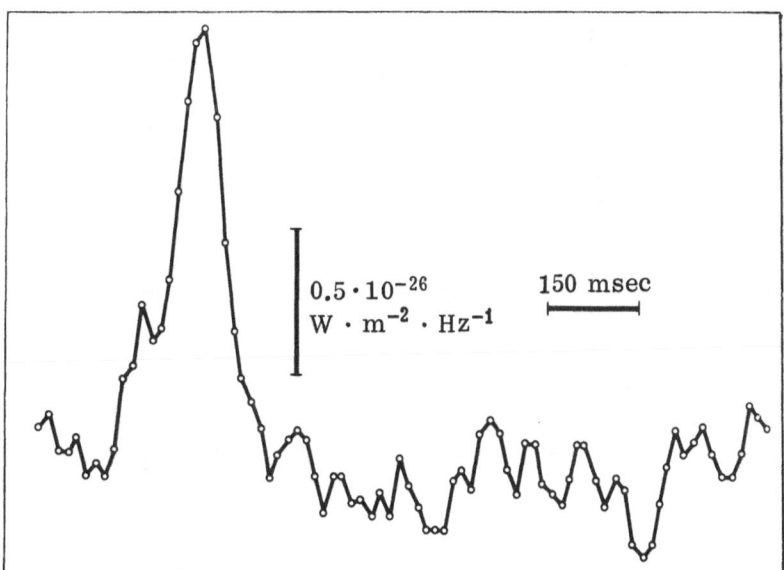

Fig. 3. Example of recording of the feeble pulsar PSR 0138 + 59. N = 172; τ = 36 msec; Δf = 800 kHz (with compensation for dispersion).

length of the waves. Therefore, in its band of frequencies, a pulse from a pulsar is washed out or blurred in time, which diminishes the ratio of signal to noise. This effect can be excluded by including in the radio receiver an arrangement for producing a dispersion equal to that produced by the interstellar medium, but opposite in sign. Practically, such a dispersion compensator can be realized by dividing the frequency band ΔF of the reception from the pulsar into a series of narrower channels Δf, the outputs of which are then integrated with the time delay

$$\Delta t_D = 0.415 \cdot 10^{16} \; MD \; (1/f_{0_n}^2 - 1/f_{0_{n-1}}^2),$$

where f_{0_n} and $f_{0_{n-1}}$ are the central frequencies of neighboring channels, and MD is the measure of dispersion of the pulsar, the shape of the pulse at the output of the receiver being determined by the dispersion only in one channel, Δf, and not over the entire band of frequencies being received, ΔF. A widening of the band in this case gives increase of sensitivity by $(\Delta F/\Delta f)^{1/2}$ times.

A simplified functional block diagram of a receiver with a dispersion compensator is shown in Fig. 4. The radio emission being received in the frequency band ΔF is divided by filters F into bands Δf, the central frequencies of which are also spaced at intervals of Δf:

$$f_{0_1} - f_{0_2} = f_{0_2} - f_{0_3} = \ldots = f_{0_{n-1}} - f_{0_n} = \Delta f.$$

The detected and amplified signals in each channel go to a delay line formed of sequentially engaged, identical elements having a delay of $\Delta \tau_d = \Delta t_d$. The amount of the dispersion of pulsar emission, determined mainly by the distance of each pulsar and characterized by the measure of dispersion, varies over a range of hundreds to one. At the mean frequency of the BSA radiotelescope of the Physics Institute ($f_0 = 102.5$ MHz) in its band of operating frequencies ($\Delta F = 2.5$ MHz), the time delay necessary for compensation (ΔF_d) between neighboring channels

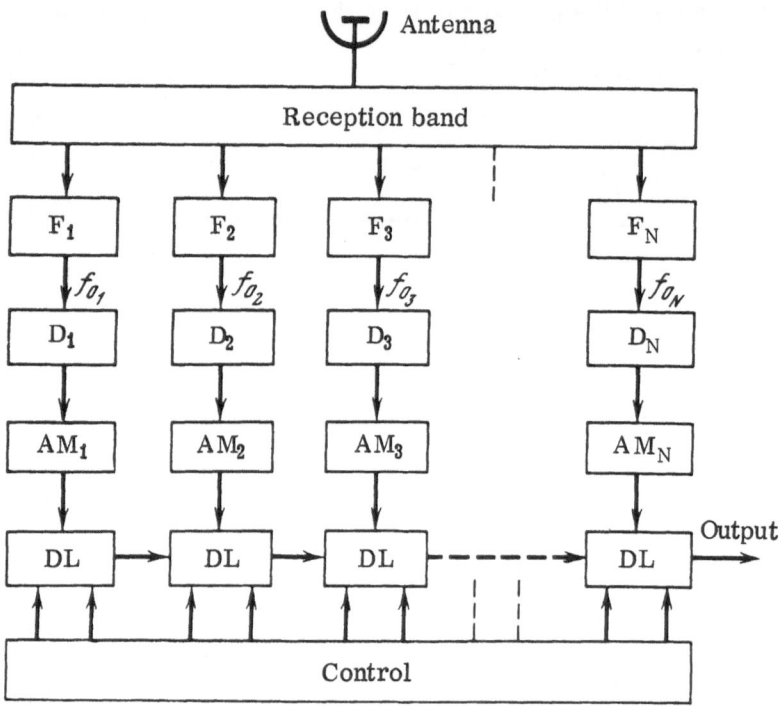

Fig. 4. Simplified block diagram of radiometer with pulsar-dispersion compensator. F, filter with a band of Δf; D, detector; AM, amplifier; DL, delay line.

varies from one to hundreds of milliseconds. Therefore the practical realization of a dispersion compensator in the form of tunable delay lines encounters great difficulties.

With the appearance of field transistors and integrating switches, it became possible to create an analog-to-digital delay circuit.

A block diagram of such a delay (Fig. 5) consists of switching (P_1 and P_2), accumulating (C_1 and C_2), and matching (K_1 and K_2) elements. At the initial moment of time, switch P_1 closes and P_2 opens. The input signal enters capacitance C_1 and charges it, and matching element K_1, having a large input resistance and a small output resistance, sustains the voltage in condenser C_1 for the time of the operation of switch P_1. Then switch P_2 closes and P_1 opens. The voltage from condenser C_1 is transmitted from the output of matching element K_1 through P_2 to storage element C_2, the action of which is sustained by matching device K_2 which has a large input resistance and small output resistance. In this circuit the delay time is determined by the difference in the moments of operation of the switching elements and can easily be changed by a change of frequency of the controlling voltage. In this way, by changing the frequency of the timing generator, we can easily "retune" or adjust the delay over a wide interval $\Delta\tau_d$.

A combination of the methods of compensation for dispersion and of synchronous accumulation of pulses makes it possible to increase the sensitivity of pulsar recording by hundreds of times and so study the average shape of the pulses with a sufficiently high time resolution.

Figure 6 shows an example of recording of the pulsar PSR 2020 + 28, with a dispersion compensator (channel II) and without the compensator (channel I), both covering the same frequency range ($\Delta F = 900$ kHz). These recordings were obtained with accumulation, or buildup, using the "Neutron" simultaneously in the two channels. It is obvious that in channel I the pulse is several times wider due to dispersion and considerably inferior as regards the signal/noise ratio.

In 1975 regular observations were begun using the new large BSA radio telescope of the Physics Institute [1] in a pulsar-search program. The radiometer's output voltage was recorded on punch tape with readout intervals of 70 msec; then the data so obtained were processed by an electronic computer. However, the punch-tape method of data recording, although it has certain merits (the possibility of repeated reproduction and the processing of the recording according to various programs and with different parameters), is practically unacceptable for a scanning search for pulsars over an entire celestial hemisphere, for which are required many months of 24-hour (or almost 24-hour) observations. The existing systems of recording on punch tape are not designed for many-hour continuous operation at the high speed of data readout (which is close to the maximum speed). The optimal way to process the long-lasting sessions of survey observations is by processing in real time, whereby the information from the radiometer's output goes directly into an electronic computer (omitting an intermediate recording in some carrier) and is processed in the course of the observations. There is also an enhancement of the operativeness or effectiveness of the processing and of the gathering of information.

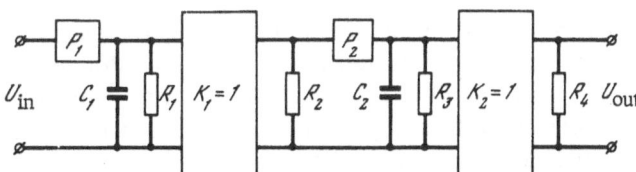

Fig. 5. Block diagram of one element of a compensator delay.

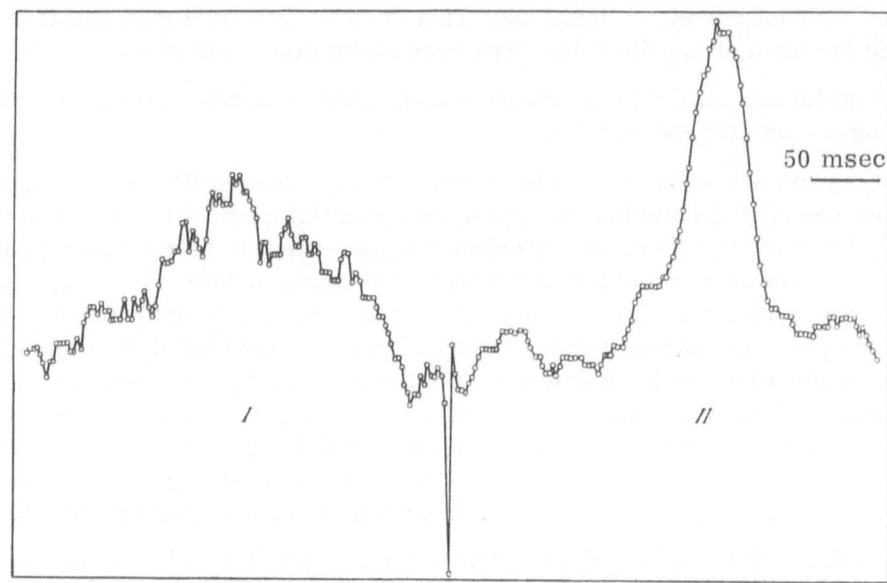

Fig. 6. Example of recording of the pulsar PSR 2020 + 28 with accumulation or buildup, using the "Neutron" analyzer over a wide reception band. I) Without compensation for dispersion; II) with compensation; $\Delta F = 900$ kHz.

The creation of such a system for processing the observations became possible in 1975 when an M-6000 electronic computer was installed at the Radio-Astronomical Station of the Physics Institute.

The basic elements of the system are the antenna, the radiometer, the dispersion compensator, the communication line, the analog-to-digital converter, the M-6000 computer complex, and the data-processing program.

Let us pause briefly to consider the system's individual units; a functional diagram of its basic elements is presented in Fig. 7. The signal from the output of a BSA (or DKR-1000) antenna goes to the input of a 12-channel radiometer. The channels, spaced in frequency, have a band width of 70 kHz each, with an interval between channels equal to the width of a channel. This 12-channel radiometer, designed for studying the fine structure of pulsar spectra, is being used in the present case for carrying out the compensation for the spectral dispersion of the pulsars' radiation as it traverses the interstellar medium, thus enhancing the sensitivity of the recording of the pulsars by 3.5 times. From the output of the compensator and before entering the electronic computer, the signal is amplified to the necessary strength by the low-frequency amplifier LFA. When recording feeble pulsars, whose individual pulses are not visible against the background of the radio telescope's intrinsic noises, the strength of the signal at the output of the amplifier is about 0.25 V (the root-mean-square value with a fixed radiometer integration time of 110 msec).

In view of the fact that the observations are being conducted with the BSA radio telescope, and the M-6000 electronic computer is installed in the RT-22 equipment house, which is situated at a distance of 700 m, a communication line has been laid between the equipment house of the DKR-1000 and the RT-22 for delivering the signal from the radiometer's output (the dispersion-compensation unit). The communication line is a 50-pair telephone cable (of the brand TB-50 × 2 × 0.5), which will from now on permit the processing of the observations in many channels, both for studying pulsars and for other programs of investigation.

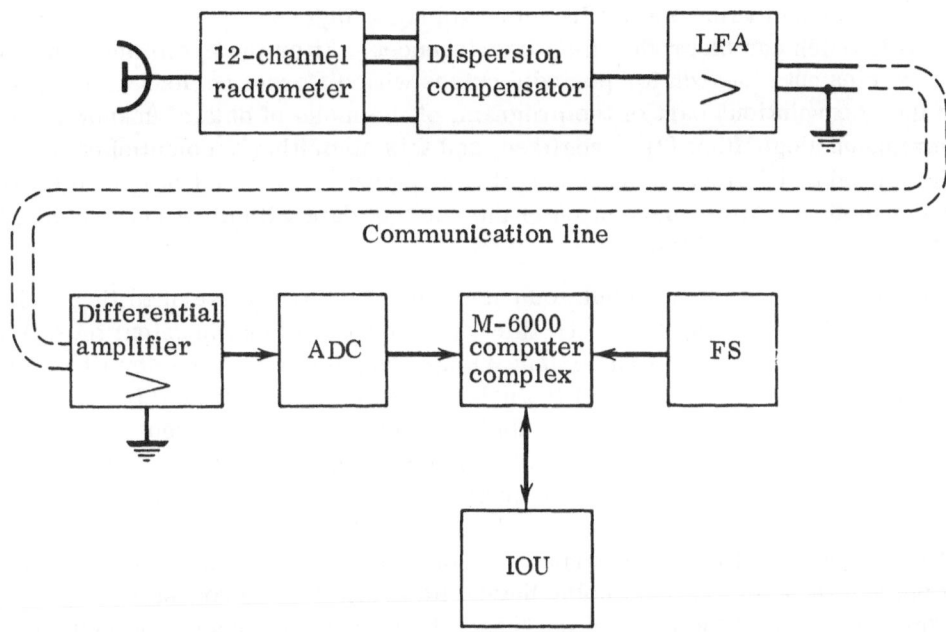

Fig. 7. Functional block diagram of the system for processing observations
of pulsars in a quasi-real-time scale.

The transmission of an analog signal to a distance involves problems caused by induction. In our case, with direct connection of the radiometer's output through the cable to the electronic computer, the induction from a 50-cycle line amounted to 5-6 V with respect to the M-6000 computer's ground. Since the induction in both conductors of a line is in phase and of approximately equal amplitude, for its suppression we installed a differential amplifier (using an IUT401V microcircuit) included between the output end of the communication line and the input of the analog-to-digital converter (ADC) of the electronic computer. The use of the differential amplifier has reduced the induction to a level of several millivolts with a magnitude of useful signal of about 1 V.

One of the basic elements of a pulsar-observation system using an electronic computer operating in real time is the program for processing the information that comes from the radio telescope. The pulsar-search program that we have set up is essentially a program for the optimal separation, from noise, of weak, periodically repeated pulses having a known period. The extraction of the signal is accomplished by the accumulation, or buildup, of the pulses from a train of as many periods as possible in a given session.

The operational procedure of the program is as follows: From a "Consul" input-output unit (IOU) or from a teletype to an M-6000 electronic computer, the required processing parameters are introduced: (1) the number of observational sessions; (2) the number of points of the mass (of input data) being processed in one session (in our case up to 4096 points); (3) the initial and final time-interval values of the periods in which pulsar search is conducted; (4) the time interval ΔT over which the data from the radiometer's output are read out and then fed into the electronic computer; (5) the threshold value of the signal (when exceeded, the amplitude of the accumulated pulse and also the period and other parameters of the accumulation are printed out). After the introduction of these parameters, imput of radiometer data into the operative memory of the computer begins. After the prescribed number of readouts, the program leads to "clearing" and preliminary processing of the recorded data, there being a computation of the parameters of the radiometer's output noise, excluding the possibility of a "flattening" of the mean value and limiting or excluding high pulse spikes (depending on the prescribed level) which can be produced by interferences. After the preliminary processing, the program carries out a search for periodic pulses with different periods in a prescribed interval. In the "convolution" part of the program, of the mass of data of unknown period, a rapidly-summing-up alogorithm [2] is realized, and this algorithm accomplishes a sorting of periods with a speed that is tens or hundreds of times greater than that of period sorting by the usual direct method. After convolution of the mass of data with different periods, the program analyzes the results so obtained.

When the amplitude of the accumulated pulse surpasses the prescribed threshold, the computer prints out the following data: the time of the observational session; the value, or duration, of the period with which a pulse appears upon accumulation; the amplitude of a pulse in units of the root-mean-square value of the noise after the accumulation; the analog value of the amplitude with respect to the input noise (before accumulation); the root-mean-square value of the input noise; the base period (auxiliary parameter) of the accumulation; and the number of pulses summed up. Also printed out are the amplitudes of the signal in all the other phases of the period, through whose values the shape of the pulsar's accumulated pulse can be plotted. If the amplitude of the signal does not surpass the prescribed threshold, then the only parameters printed out are those of the input noise before its preliminary processing and after "clearing." Since the processes of the input of the mass of data and its subsequent analysis are separated in time, we call this processing a search in quasi-real-time scale.

Let us dwell briefly on some of the program's characteristics. The processing program is written in the Fortran-II language, with the exception of the subprogram for the input of data from the analog-to-digital converter into the computer's operating memory, which is written in assembler (mnemocode). It consists of a basic block and seven subprograms, and together with the library subprograms it occupies 5500 memory elements. The processing of search observations in quasi-real-time scale became possible thanks to the sufficiently great speed of operation of this program. Thus, for example, one of the observation sessions, which lasted 4 min 06 sec (4096 readouts with intervals of 60 msec), was processed in 2 min 46 sec, the search being conducted at intervals of 0.24 to 1.02 sec, within whose limits are found the periods of a majority (70%) of pulsars.

This system of processing in quasi-real-time scale was tested in August, 1975, with observations of the known pulsars CP 0809, CP 0834, MP 1508, and PP 0943, from which the expected results were obtained. A copy of part of the M-6000 computer's processing with a

TABLE 1. Pulsar Search Program (FFA)

IMAX: 4096
SAMPL: .034
LIMIT: 5.0
RANGE: 8—32

NOB	AMEAN	RMS	RMSC	ALRMS	TIME	PERIOD	PULS	FLUX	RMSN	MP	NP
1	—15.9	15.36	10.23	10.07							
					9.43.23	.548435	6.0	.38	10.07	16	256
—2.6	—2.6	—2.0	—.5	.2	4.1	6.0	3.6	2.2	1.6	.6	—1.1
—2.3	—3.6	—2.7	—1.6								
					9.43.23	1.096905	6.7	.59	10.07	32	128
					9.43.23	1.096905	9.0	.80	10.07	32	128
					9.43.23	1.096905	6.2	.55	10.07	32	128
—1.4	—1.3	—1.3	—.3	.9	6.7	9.0	6.2	3.5	2.0	.5	—1.5
—2.6	—3.0	—2.4	—1.5	—2.3	—2.3	—1.6	—.4	—.7	—.9	—.5	—1.0
—.4	.3	.3	—.0	—.7	—1.3	—1.3	—.8				

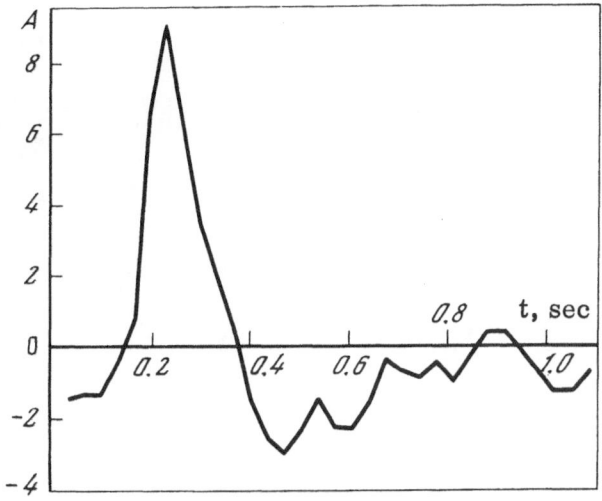

Fig. 8. Mean profile of pulse from the pulsar PP 0943, obtained with the help of an M-6000 electronic computer.

pulsar search program in quasi-real-time scale while observing the pulsar PP 0943 is shown in Table 1. With 4096 readouts and 34 msec between readouts, analysis was conducted within the limits of periods from 0.272 to 1.122 sec (8 to 32 by points). As a result of the processing the electronic computer "discovered" the pulsar with a period of 1.0969 sec and amplitude 9 with respect to the root-mean-square value σ of the noise after accumulation (0.8 with respect to the input noise), 128 pulses from the pulsar PP 0943 being accumulated. Since the amplitude of accumulated pulse was nearly twice that of the threshold (5σ), then naturally a pulse (Fig. 8) appeared with an accumulation period double the shortest one (0.5484 sec) but with a smaller amplitude (6). It should be noted that the value obtained for PP 0943's period is very near its tabular value of 1.0977 sec.

As a result of these observations, the pulsar's mean flux can be estimated. The root-mean-square value of the input-noise flux σ_{in} in these observations was 1 flux unit. Consequently, the mean flux at the pulse's maximum was $0.8 \cdot 10^{-26}\, W \cdot m^{-2} \cdot Hz^{-1}$. From the shape of the mean profile (see Fig. 8), we can compute the flux averaged over the entire interval of the pulsar, which for PP 0943 is $\simeq 0.1$ units.

The application of digital computing technology to the processing of the results of observation in real time has increased by many times the sensitivity and effectiveness of pulsar research conducted with the DKR-1000 and BSA radio telescopes of the Physics Institute. At the present time, with the help of this technique, a survey of known pulsars is being successfully conducted over the northern stellar hemisphere at 102.5 MHz (with the BSA), and already the first interesting results have been obtained. Spectral measurements have been made of more than ten pulsars in a range of 60-110 MHz. A search for new pulsars with computer processing in quasi-real-time scale has been initiated.

LITERATURE CITED

1.   V. V. Vitkevich, A. A. Glushaev, Yu. P. Ilyasov, S. M. Kutuzov, A. D. Kuz'min, M. M. Tyaptin, I. A. Alekseev, V. N. Brezgunov, V. F. Bunin, G. F. Novozhenov, G. A. Pavlov, G. S. Podlipnoi, and N. S. Solomin, The BSA Antenna-Aperture Complex of the Physics Institute of the Academy of Sciences of the USSR. Report at the 8th All-Union Conference on Radio Astronomy, Pushchino (1975).
2.   Yu. P. Shimov, Astron. Zh., 49:470 (1972).
3.   D. H. Staelin, Proc. IEEE, 57:724 (1969).

PROGRESS IN RADIO-ASTRONOMIC INVESTIGATIONS OF THE SOLAR-WIND VELOCITY

N. A. Lotova and I. V. Chashei

Various methods for the correlation and spectral analysis of drift processes are discussed, along with their capabilities in application to interplanetary scintillations. The influence of fine structure in the solar-wind velocities and time variation of inhomogeneities of the medium on the parameters of the scintillation pattern are analyzed. The theory developed in the article is used as a basis for analysis of the experimental data. It is shown that the presence of velocity structure along the line of sight can be fully explained by rearrangement of the scintillation pattern at distances from the sun $\rho \geq 0.5$ a.u.

INTRODUCTION

The radio-astronomic method of investigating a medium by means of the scintillations of radio sources was proposed by Ginzburg in 1956 [1]. Interplanetary scintillations (IPS) were discovered in 1964 [2], and in the ensuing years the method has been used extensively for solar-wind studies. The scintillation method has now emerged as a popular means of investigating the interplanetary [3-12], interstellar [13], and ionospheric [14, 15] plasma. It is used to study the inhomogeneous component of the medium, including the parameters of the inhomogeneities, their dynamics and movements, and their distribution in space.

The scintillation method is essentially based on the fact that radiation traveling from compact radio sources through a medium containing inhomogeneities of the refractive index suffers diffraction, producing spatial intensity fluctuations on earth. In this case the observer perceives the spatial intensity fluctuations of translucent sources as temporal fluctuations, or scintillations. By observing the drift of the diffraction pattern on earth it is possible to determine the velocity vector of the pattern from the time lag of the scintillations between two or three stations.

The present study is concerned primarily with IPS, but the fundamental results of the theory are equally applicable to interstellar and ionospheric scintillations. The interpretation of the scintillations was originally based on the assumption that the diffraction pattern is "frozen," implying that the inhomogeneities do not change with time and are "frozen into" the interplanetary plasma, the medium moving as an integral whole with a certain time-constant velocity V_0 relative to the line of sight. It is important to note that if the frozen-pattern hypothesis is true, a one-to-one correspondence will exist between the spatial spectrum of the inhomogeneities and the temporal spectrum of the scintillations [16, 17]. On the basis of physical considerations, however, it is unreasonable to expect all inhomogeneities in the medium to have precisely the same velocity. The occurrence of different velocities in the medium, along with temporal variation of the inhomogeneities, has the effect that the temporal intensity fluctuations are not the result of simple translation of the spatial fluctuations. This means that

85

the motion of the diffraction pattern is accompanied by internal rearrangement. The pattern-rearrangement effect of IPS was discovered fairly recently [18]. The effect is normally described in the language of random velocities in the diffraction pattern. The occurrence of rearrangement indicates that the frozen-pattern hypothesis is not realized in general and is in fact a rather crude approximation. Considerable importance attaches, therefore, to the problem of analyzing the random velocities in the observed IPS pattern. This problem is significant insofar as the study of random velocities in the diffraction pattern makes it possible to analyze a number of other problems associated with the physics of the solar wind and its inhomogeneous component. Foremost among these problems is the nature of the rearrangement of the diffraction pattern, i.e., whether the random velocities are affiliated with dispersion of the velocities of the inhomogeneities themselves or whether they are attributable to actual variation of the inhomogeneities with time. Also, for reconstruction of the total flow pattern of matter from the sun it is important to know how the structure of the velocities in the solar wind is distributed and how it is related to other parameters of the solar wind. Finally, it is necessary to consider the random velocities in the diffraction pattern in order to reconstruct the spatial spectrum of the inhomogeneities by means of the observable temporal scintillation spectrum. Thus, the investigation of the solar wind velocity and its fine structure provides a means for studying the physical processes occurring on the sun and in the interplanetary medium.

Currently there are two approaches to the analysis of diffraction-pattern drifts: the cross-correlation and cross-spectral methods. Several different techniques have been developed within the context of these approaches for studying the velocity of the inhomogeneities and its structure in the interplanetary medium. The exposition that follows is given over to a critical analysis of the existing methods and a comparative study of their capabilities. In the last section of this article we analyze the experimental data on IPS.

1. Methods of Cross-correlation Analysis
of Interplanetary Scintillations

The intensity of radio radiation transmitted through a medium with fluctuations of the refractive index is represented in the pattern plane (x, y) by a stochastic function of the coordinates and time $I(\mathbf{r}, t)$. The temporal variations of the radiation intensity I at point $\mathbf{r}$ can be attributed to two causes: (1) motion of the spatial diffraction pattern of $I(\mathbf{r})$ with the average velocity of the scattering flow; (2) internal rearrangement of the pattern as it travels. The latter circumstance can be associated both with the presence of physically different velocities along the line of sight (wind pattern) and with temporal variation of the inhomogeneities (vorticity pattern). The rearrangement of the scintillation pattern is usually discussed in terms of the wind pattern, in which case we speak of the distribution function of the inhomogeneity velocities $\varphi(\mathbf{V})$ [the statistical dispersion of the velocities for the distribution function $\varphi(\mathbf{V})$ is measured by the variance of the velocity or its square root, the standard deviation (SD)]. Here we abide by this interpretation without attempting to justify, for the time being, the hypothesis of the wind character of the IPS pattern, although, as will become apparent later, there are plausible reasons to believe that the governing contribution to rearrangement of the IPS pattern comes from the occurrence of physically different velocities along the line of sight.

1.1. Old Methods of Cross-correlation Analysis

Suppose that the intensity fluctuations $\delta I(\mathbf{r}_i, t)$ of a radio source are recorded simultaneously at several different stations with coordinates r_i ($i = 1, 2, \ldots$). Cross-correlation analysis entails investigation of the cross-correlation functions $B_I(\mathbf{r}_{ij}, \tau) = \langle \delta I(\mathbf{r}_i, t)\delta I(\mathbf{r}_j, t + \tau)\rangle$ and autocorrelation functions ($i = j$) of the intensity fluctuations. The objective of cross-correlation analysis is to determine the spatial correlation function of the intensity fluctuations

$B(\mathbf{r}) = \langle \delta I(\mathbf{r_1}) \, \delta I(\mathbf{r_1 + r}) \rangle$ (in particular, the space scale of the diffraction pattern), the average velocity vector $\langle \mathbf{V} \rangle$ of the diffraction pattern, and the velocity standard deviation (SD) $\sigma[\langle \mathbf{V} \rangle = \int \mathbf{V} \varphi(\mathbf{V}) d\mathbf{V}, \; \sigma^2 = \int (\mathbf{V} - \langle \mathbf{V} \rangle)^2 \varphi(\mathbf{V}) d\mathbf{V}$, where $\varphi(\mathbf{V})$ is the velocity distribution function]. We now consider various drift-analysis methods developed in papers on the cross-correlation approach.

a. Method of Total Correlation Analysis (BPS Method). This method was first proposed by Briggs, Phillips, and Shinn [19] and was used in the analysis of ionospheric and interplanetary scintillations. We discuss the basic ideas of the BPS method in the one-dimensional modification.

Suppose that an observer is situated at a large number of points (stations) along the OX direction, which coincides with the velocity vector of the scattering flow. If the intensity is measured simultaneously at all stations at two different times t and t + τ' (i.e., if two instantaneous "photographs" of the spatial pattern are obtained), it is then possible to form the spatial cross-correlation function $\rho(x, \tau')$ and the autocorrelation function $\rho(x, 0)$:

$$\rho(x, \tau') = \frac{1}{\langle \delta I^2 \rangle} \langle \delta I(x_1, t) \, \delta I(x_1 + x, t + \tau') \rangle,$$
$$\rho(x, 0) = \frac{1}{\langle \delta I^2 \rangle} \langle \delta I(x_1, t) \, \delta I(x_1 + x, t) \rangle. \tag{1}$$

The autocorrelation function $\rho(x)$ represents the sought-after spatial correlation of the scintillations. Due to the motion of the medium with average velocity $\langle \mathbf{V} \rangle$ the pattern moves in the x direction, and so the function $\rho(x, \tau')$ attains a maximum at some $x = x' \neq 0$. The velocity $\langle \mathbf{V} \rangle$ of the scintillation pattern is defined as the velocity of the frame in which the variations in the pattern are a minimum. Knowing the position of the maximum of $\rho(x, \tau')$, we can determine the velocity $\langle V \rangle = x'/\tau'$.

The procedure is readily extended to the two-dimensional case. If observations are made at a large number of stations situated in all possible directions, then the spatial autocorrelation function $\rho(\mathbf{r})$ of the scintillations can be obtained, and the velocity $\langle \mathbf{V} \rangle$ determined by comparison of the two-dimensional patterns of $\delta I(\mathbf{r}, t)$ at times t and t + τ'. The time τ' occurs in the function $\rho(\mathbf{r}, \tau')$ as a parameter, hence we cannot assess the random velocities in the diffraction pattern by comparing two instantaneous "photographs."

Suppose now that the temporal intensity fluctuations are recorded at only two stations spaced a distance x apart. The problem of determining the one-dimensional spatial correlation $\rho(x)$ and average velocity $\langle \mathbf{V} \rangle$ becomes indeterminate because of the temporal rearrangement of the scintillation pattern. Here the form of the temporal autocorrelation function $\rho(\tau)$ of the scintillations coincides with the form of the spatial function $\rho(x)$ only in the event that there is no spread of the velocities (SD of the velocities $\sigma = 0$). Temporal rearrangement of the pattern can produce an appreciable disparity between the form of $\rho(\tau)$ and the form of $\rho(x)$ [20]. The temporal cross-correlation function $\rho(x, \tau)$ of the scintillations attains a maximum at some $\tau = \tau_0$. The velocity V_a determined at the position τ_0 of the maximum, namely $V_a = x/\tau_0$, will differ from the true velocity $\langle \mathbf{V} \rangle$, $V_a \geq |\langle \mathbf{V} \rangle|$, due to pattern rearrangement, so that V_a is usually called the apparent velocity.

In a more general setting we can discuss the two-dimensional correlation surface $\rho(x, \tau)$. It follows from the definition of the function $\rho(x, \tau)$ (1) that it is bounded, $|\rho(x, \tau)| \leq 1$, and even, $\rho(x, \tau) = \rho(-x, -\tau)$. This means that $\rho(x, \tau)$ describes a central surface with center at the point $x = \tau = 0$ and the argument of the function $\rho(u)$ contains only even powers of x and τ. Moreover, we infer from the centrality of the surface $\rho(u)$ that the constant-correlation contours u = const are closed. It is customarily assumed that $\rho(u)$ is a Gaussian-type function

$$\rho(u) = \rho(x, \tau) = \rho(a_{11}x^2 + 2a_{12}x\tau + a_{22}\tau^2) \tag{2}$$

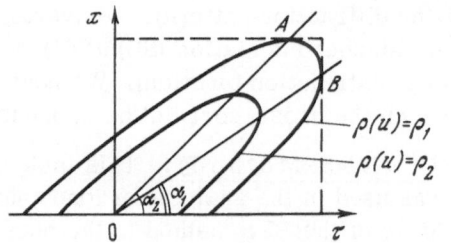

Fig. 1. Constant-correlation contours $\rho(x, \tau) =$ const in the (x, τ) plane.

and, as such, adequately describes the investigated drift process. Under this assumption the quadratic form in the argument of (2) is positive definite, and the constant-correlation contour is an ellipse*:

$$a_{11}x^2 + 2a_{12}x\tau + a_{22}\tau^2 = \text{const.} \tag{3}$$

The constant-correlation ellipses (3) are concentric and, upon transformation to a frame moving with the drift velocity $\langle V \rangle$ $(x_1 = x - \langle V \rangle \tau)$ are reduced to the principal axes:

$$x_1^2/a^2 + \tau^2/\tau_c^2 = \text{const,} \tag{4}$$

where a is a characteristic space scale of the diffraction pattern and τ_c is the relaxation or rearrangement time, which is related to the velocity SD σ by the equation $\tau_c = a/\sigma$.

The correlation function $\rho(x, \tau)$ (2) with one argument fixed is a maximum for $\partial u/\partial \tau = 0$ if $x = \text{const}$ (Fig. 1, point A), or for $\partial u/\partial x = 0$ if $\tau = \text{const}$ (Fig. 1, point B). In the first case the maximum is attained at $\tau = \tau_0 = x/V_a$ $(x = \text{const})$, and in the second case at $x = x' = \langle V \rangle \tau'$ $(\tau = \text{const})$. From this result we infer that, knowing the constant-correlation ellipse (see Fig. 1), we can determine the velocities $\langle V \rangle$ and V_a: $\langle V \rangle = \tan \alpha_1$, $V_a = \tan \alpha_2$. The reduction of the ellipse (3) to the principal axes (4) enables us to determine the correlation space scale a. The apparent velocity V_a of the pattern is related to the drift velocity $\langle V \rangle$ by the expression [19, 20]

$$V_a = \langle V \rangle + \sigma^2/\langle V \rangle, \tag{5}$$

which can be used to determine the velocity SD σ.

An important attribute of the constant-correlation contour (3) (see Fig. 1) is the symmetry of the points of intersection of the ellipse with the lines $x = \text{const}$ and $\tau = \text{const}$ about the respective lines OA and OB.

Thus, to obtain the desired characteristics of the IPS pattern, namely the correlation scale a, drift velocity $\langle V \rangle$, and velocity SD σ, in the BPS method we construct the correlation surface $\rho(x, \tau)$ and from it determine the constant-correlation ellipses (see Fig. 1).

When the scintillation method is used to study the solar wind, we obtain two correlation cross sections from experiment: $\rho(x = \text{const}, \tau)$ and $\rho(0, \tau)$. Using these cross sections and specifying the form of the spatial correlation function $\rho(x)$ with one free parameter (correlation scale), we can construct the constant-correlation ellipse (Fig. 1). However, difficulties are met in application of the BPS method, first, because the number of stations used is limited, usually to not more than three [21], and, second, because the observation stations are not generally located along the flow velocity, but at the points of a certain triangle (often almost equilat-

---

* The inclusion of higher powers of x and τ is disussed later (see Section 1.3) and has the effect of exposing certain subtle details of the function $\rho(x, \tau)$.

eral), and the orientation of the velocity $\langle \mathbf{V} \rangle$ relative to the triangle varies. The first difficulty is fundamental to radio astronomy and makes it impossible to determine the spatial correlation $\rho(\mathbf{x})$. Accordingly, the form of the function $\rho(\mathbf{x})$ is specified a priori, and only the parameters of $\rho(\mathbf{x})$ (correlation scale) are determined. The second difficulty makes it necessary to invoke the two-dimensional modification of the method and complicates the experimental data-processing procedure accordingly. The BPS method is generalized to the two-dimensional case in [22], making it possible, once the form of the function $\rho(\mathbf{r})$ has been specified, to determine the correlation scale of the intensity fluctuations, the drift velocity vector $\langle \mathbf{V} \rangle$, and the random velocity SDs σ_x and σ_y along the coordinate axes. This procedure is based on construction of the characteristic velocity ellipse and then determination of the velocity $\langle \mathbf{V} \rangle$ from the front of the apparent velocities. The BPS method is generalized to the three-dimensional case in [23, 24].

The BPS method requires ellipticity on the part of the constant-correlation contours, i.e., it is necessary that the function $\rho(\mathbf{r}, \tau)$ depend quadratically on the space and time variables. This requirement imposes rather stringent constraints on the usability of the method. First, a quadratic correlation function occurs for rather small shifts of $|\mathbf{r}|$ and τ, so that the BPS works well only for baselines $|\mathbf{r}| \ll a$ (a is the correlation scale) and time shifts $\tau \ll a/\langle \mathbf{V} \rangle$. In the investigation of IPS the usual baselines [21] are $r \sim a$, thus depreciating considerably the reliability of the method. Second, the space and time variables enter into the argument of $\rho(\mathbf{r}, \tau)$ in like powers only for a small velocity SD $\sigma \ll \langle \mathbf{V} \rangle$. The velocity SD σ attains 30% in the interplanetary medium, further degrading the capabilities of the method.

    b. **Method of Intersections.** This method was proposed by Armstrong and Coles [18]. It postulates a quadratic argument of the spatial correlation function $\rho(\mathbf{r}) = \rho[Q(\mathbf{r})]$, where $Q(\mathbf{r}) = \mathbf{r} A \mathbf{r}$ is a quadratic form and A is a symmetric matrix. It is also postulated that the temporal cross-correlation function obtained at two stations separated by a distance $\mathbf{r}$ has the form $\rho(\mathbf{r}, \tau) = \rho[Q(\mathbf{r} - \langle \mathbf{V} \rangle \tau), \tau]$.

The essence of the method is as follows. The temporal correlation functions obtained on different baselines $\mathbf{r}_i \neq \mathbf{r}_j$ intersect at certain $\tau = \tau_{ij}$, with the intersection condition $\rho(\mathbf{r}_i, \tau_{ij}) = \rho(\mathbf{r}_j, \tau_{ij})$. The intersections τ_{ij} exist for all baseline pairs $\mathbf{r}_j, \mathbf{r}_i$, including $\mathbf{r}_i = 0$ for the autocorrelation. Inasmuch as $\rho(\mathbf{r}_i, \tau_{ij}) = \rho(\mathbf{r}_j, \tau_{ij})$ for $Q(\mathbf{r}_i - \langle \mathbf{V} \rangle \tau_{ij}) = Q(\mathbf{r}_j - \langle \mathbf{V} \rangle \tau_{ij})$, we have

$$\langle \mathbf{V} \rangle A (\mathbf{r}_i - \mathbf{r}_j) = \frac{1}{2\tau_{ij}} [Q(\mathbf{r}_i) - Q(\mathbf{r}_j)].$$

Thus, for any two nondegenerate intersections ($\tau_{ij} \neq 0$) it is possible to obtain the velocity vector $\langle \mathbf{V} \rangle$. It is important to note that for any given baseline (except $\mathbf{r}_j = 0$) $\rho(-\mathbf{r}_i, \tau) = \rho(\mathbf{r}_i, -\tau)$, and so additional intersections can be used. The matrix A can be deduced from the values of the zero-shift correlation function $\rho(\mathbf{r}_i, 0)$. In determining the velocity $\langle \mathbf{V} \rangle$ from the intersections τ_{ij} any scalar factor in $Q(\mathbf{r})$ drops out, so that in the determination of $\langle \mathbf{V} \rangle$ it is not necessary to know the spatial correlation scale of the scintillations, rather it is sufficient to have the orientations of the principal axes of the matrix A and the ratio of the correlation scales along the principal axes. From the values found for the velocity $\langle \mathbf{V} \rangle$ it is possible to determine the spatial correlation scale a and velocity Sd σ. For this operation it is necessary to use the relations for the apparent velocity ∇_a and the apparent scale a_a of the diffraction pattern [20] *

$$V_a = \langle V \rangle \left(1 + 2 \frac{\sigma^2}{\langle V \rangle^2} \right); \qquad a_a = a \left(1 + 2 \frac{\sigma^2}{\langle V \rangle^2} \right)^{1/2}. \tag{6}$$

---

*Relations (6) have been obtained for a two-dimensional isotropic Gaussian velocity distribution; their derivation is given in § 1.3.

Consequently, the method of intersections enables us to solve the correlation analysis problem, i.e., to determine the velocity $\langle \mathbf{V} \rangle$, the scintillation scale a, and the velocity SD σ, with milder requirements on the form of the space-time correlation function $\rho(\mathbf{r}, \tau)$ than in the BPS method; the only requirement is that the combination $(\mathbf{r} - \langle \mathbf{V} \rangle \tau)$ enter into the argument of $\rho(\mathbf{r}, \tau)$ as a quadratic variable, but it is not required that the dependence of $\rho(\mathbf{r}, \tau)$ on the space and time variables have the same form. The intersection method is simpler than the BPS method and has one additional advantage over the latter: In determining the velocity $\langle \mathbf{V} \rangle$ it is not necessary to take into consideration rearrangement of the IPS pattern. Nonetheless, the requirement that the correlation function $\rho(\mathbf{r}, \tau)$ have the form $\rho(\mathbf{r} - \langle \mathbf{V} \rangle \tau, \tau)$, is still fairly strong and is upheld only when the deviations of the velocities of the inhomogeneties from the average $\langle \mathbf{V} \rangle$ are isotropically distributed in space. Therefore, all the difficulties incurred by the BPS method in analysis of the velocities are essentially preserved by the intersection method.

c. Little – Ekers (LE) Method. The LE method is described in [25]. Unlike the BPS and intersection methods, here it is not demanded that the constant-correlation contours be ellipses (BPS) or that the function $\rho(\mathbf{r}, \tau)$ depend quadratically on $(\mathbf{r} - \langle \mathbf{V} \rangle \tau)$ (intersections).

In the LE Method a definite form (usually Gaussian) is assigned to the spatial correlation function $\rho(\mathbf{r})$ of the IPS. From the known values of the cross-correlation $\rho(\mathbf{r}_i, 0)$ for zero shift τ we determine the correlation scales, which are parameters of the function $\rho(\mathbf{r})$. We then specify the velocity distribution function $\varphi(\mathbf{V})$ with free parameters, namely the projections $\langle \mathbf{V} \rangle_x$, $\langle \mathbf{V} \rangle_y$ of the average velocity and the SD of the velocities σ. The correlation function for uniform motion $\rho(\mathbf{r} - \mathbf{V} \tau)$ is averaged over the velocity distribution $\varphi(\mathbf{V})$. Here the parameters $\langle \mathbf{V} \rangle_x$, $\langle \mathbf{V} \rangle_y$, σ are varied in such a way as to best fit the calculated function $\rho(\mathbf{r}, \tau) = \int \rho(\mathbf{r} - \mathbf{V} \tau) \times \varphi(\mathbf{V}) d\mathbf{V}$ into the experimentally determined cross-correlation function $\rho_{exp}(\mathbf{r}, \tau)$. Thus, the LE method makes it possible to obtain the average drift velocity $\langle \mathbf{V} \rangle$ and the velocity SD σ from the given spatial correlation $\rho(\mathbf{r})$ and velocity distribution function $\varphi(\mathbf{V})$. One difficulty of this method is the fact that its application requires a priori specification of the spatial correlation function $\rho(\mathbf{r})$ and velocity distribution $\varphi(\mathbf{V})$.

All three of the foregoing methods have been used for analysis of the IPS pattern [18]. The magnitudes of the solar-wind velocities $\langle \mathbf{V} \rangle$ obtained by these methods have turned out to be in good agreement with each other, but roughly one-third the value of the apparent velocity V_a. On the other hand, V_a agrees well with measurements of the velocity of the proton component of the solar wind from on board the Vela-3.5 satellites. The reason for this discrepancy is not yet clear, but it apparently mirrors the above-stated difficulties of the methods used for correlation analysis of the scintillations. Thus, such a marked disparity between the apparent velocity V_a and the average velocity $\langle \mathbf{V} \rangle$ cannot be attributed to the presence of strong velocity dispersion, because in that event the two-station correlation of the scintillations would be inconsequential as a result of pattern rearrangement. In any case, the fact remains that the BPS, LE, and intersection methods yield far too low a value for the solar-wind velocity, and so in the interpretation of the scintillations it is more sensible to determine the velocity of the inhomogeneities by another method, which we now consider.

1.2. Form of the Correlation Function: Statement
of the Problem

The investigation of the solar wind by the scintillation method dates back to 1964 and mainly involves the determination of effective overall parameters along the line of sight, namely the rms density fluctuations and the average velocity and characteristic scales of the inhomogeneities. In the present phase we are witnessing a transition from the investigation of overall parameters to more subtle effects typifying the velocity and density distributions of the solar wind along the line of sight. Of major importance are problems associated with the fine structure (spread) of the velocities of the inhomogeneities, namely direct estimation of the standard deviation (SD) of the velocities and the explication of its nature, investigation of the dependence of the velocity fine structure on the elongation and heliolatitude, study of the relationship between the temporal variations of the velocity fine structure and the variations of other parameters of the solar wind, and determination of the form of the spatial spectrum of the inhomogeneities from the temporal spectrum of the scintillations with allowance for the velocity fine structure. Knowledge of the spatial spectrum is important in this connection for understanding the nature of the solar-wind inhomogeneities, while knowledge of the nature of the velocity fine structure of the solar wind and the peculiarities of its spatial distribution provides a means for assessing the physical processes taking place on the sun. To solve this category of problems it is necessary to have access to a method for analysis of the velocity fine structure. This consideration has prompted the submission and discussion of the question as to how the presence of velocity fine structure is felt in the form of the temporal cross-correlation function $\rho(\mathbf{r}, \tau)$ of the scintillations, along with its spectrum $P_c(\omega)$ (see Section 2), or, in other words, how the velocity fine structure can be detected and its characteristic parameters described on the basis of the known IPS cross-correlation functions.

Below we give the fundamental relations supporting the indicated anaysis. The principal hypothesis underlying the calculations is the wind nature of the velocity structure. If velocity spread is absent, the IPS pattern is not rearranged during its motion, and the form of the temporal cross-correlation function $\rho(\mathbf{r}, \tau)$ coincides with the form of the temporal and spatial autocorrelation functions $\rho(0, \tau)$ and $\rho(\mathbf{r}, 0)$. It is assumed that rearrangement of the IPS pattern does occur and is determined by the presence of physically different velocities along the line of sight (variation of the inhomogeneities are ignored here), and that the velocities of the inhomogeneities fit a certain distribution function $\varphi(\mathbf{V})$. We consider the correlation function $\rho(\mathbf{r}, \tau)$ of the intensity fluctuations of radiation transmitted through such a medium. In the case of weak scintillations, such that the rms phase-angle lead of the scattering-inhomogeneity wave $\Phi_0 \ll 1$, the presence of unequal velocities in the medium results in independent summation of the local uniform motions. The cross-correlation function of the intensity fluctuations in this case can be written in the form [20, 25]

$$\rho(\mathbf{r}, \tau) = \int \rho^{(0)}(\mathbf{r} - \mathbf{V}\tau)\varphi(\mathbf{V})d\mathbf{V}, \tag{7}$$

provided that the statistical inhomogeneity of the medium is small-scale. In the case of saturated scintillations ($\Phi_0^2 \gtrsim 1$) the correlation function $\rho(\mathbf{r}, \tau)$ is [26]

$$\rho(\mathbf{r}, \tau) = \exp[-D_\Phi(\mathbf{r}, \tau)],$$
$$D_\Phi(\mathbf{r}, \tau) = 2\Phi_0^2\left[1 - \int \rho_\Phi^{(0)}(\mathbf{r} - \mathbf{V}\tau)\varphi(\mathbf{V})d\mathbf{V}\right], \tag{8}$$

where $D_\Phi(\mathbf{r}, \tau)$ and $\rho_\Phi^0(\mathbf{r})$ are the structural and local correlation functions of the wave phase fluctuations.

We can use relations (7) and (8) to exhibit the singular attributes of the function $\rho(\mathbf{r}, \tau)$ due to the presence of fine structure in the solar-wind velocities. We give particular attention to the weak-scintillation case (7), because it yields more information in the investigation of the solar wind; the relationship between the correlation function of the intensity fluctuations and the correlation function of the refractive-index fluctuations of the medium is linear in this case [27-31]. As an example we discuss a normal distribution of velocities $\varphi(\mathbf{V})$ in a medium with isotropic dispersion $\sigma_x = \sigma_y = \sigma$:

$$\varphi(\mathbf{V}) = \frac{1}{2\pi\sigma^2}\exp\left[-\frac{(\mathbf{V} - \langle V \rangle)^2}{2\sigma^2}\right] \tag{9}$$

and a Gaussian spatial correlation function for the scintillations, which implies that the medium is characterized by a separate inhomogeneity space scale a:

$$\rho\,(\mathbf{r}) = \exp\,[-r^2/2a^2]. \tag{10}$$

The substitution of (9) and (10) into (7) yields the following expression for the space-time correlation $\rho(\mathbf{r}, \tau)$ [20]:

$$\rho\,(\mathbf{r}, \tau) = \left(1 + \frac{\tau^2}{\tau_c^2}\right)^{-1} \exp\left[- \frac{(\mathbf{r} - \langle V \rangle\,\tau)^2}{2a^2\,(1 + \tau^2/\tau_c^2)}\right], \tag{11}$$

where $\tau_c = a/\sigma$ is the already-indicated relaxation (rearrangement) time of the IPS pattern. Expression (11) illustrates the dependence of the form of the IPS cross-correlation function on the velocity SD σ in the scattering medium and can be used to determine σ. It is evident from (11) that for small velocity SD ($\sigma \ll |\langle V \rangle|$) the BPS method is applicable only for $\tau \ll \tau_c$, because for $\tau \sim \tau_c$ the constant-correlation contours lose their ellipticity. But if the velocity SD is large enough ($\sigma \gtrsim |\langle V \rangle|$), the BPS method is totally inapplicable for analysis of the velocity structure.

1.3. Parameters of the Scintillation Cross-correlation

Function (Weak Scintillations)

a. Drift Velocity of Diffraction Pattern. Taking the fine structure of the velocities $\varphi(\mathbf{V})$ into consideration, we determine the diffraction-pattern velocity, which of course is obtained from the maximum of the cross-correlation function $\rho(\mathbf{r}, \tau)$ (7) in spatial (τ = const) and temporal ($\mathbf{r}$ = const) measurements of the intensity fluctuations.

We first consider the simple model of a medium with two characteristic velocities $\mathbf{V_1}$ and $\mathbf{V_2}$, which can be realized, for example, with an accelerated flow present along the line of sight [32-34]. We assume that the velocities $\mathbf{V_1}$ and $\mathbf{V_2}$ correspond to scintillation indices m_1 and m_2. We also assume that the spatial correlation functions $\rho_{1,2}(\mathbf{r})$ of the intensity fluctuations have a Gaussian form:

$$\rho_{1,\,2}\,(\mathbf{r}) = m_{1,\,2}^2 \exp\left(- \frac{r^2}{2a_{1,\,2}^2}\right). \tag{12}$$

Then the correlation function $\rho(\mathbf{r}, \tau)$ assumes the form

$$\rho\,(\mathbf{r}, \tau) = m_1^2 \exp\left[- \frac{(\mathbf{r} - \mathbf{V_1}\tau)^2}{2a_1^2}\right] + m_2^2 \exp\left[- \frac{(\mathbf{r} - \mathbf{V_2}\tau)^2}{2a_2^2}\right]. \tag{13}$$

The scintillation-pattern drift velocity observed on earth for the spatial fluctuations (τ = const) is determined from the condition

$$\left.\frac{\partial \rho}{\partial r}\right|_{\tau=\text{const}} = 0 \tag{14}$$

and on the basis of (13) is equal to

$$\langle V \rangle = \frac{(m_1^2/a_1^2)\,\mathbf{V_1} + (m_2^2/a_2^2)\,\mathbf{V_2}}{m_1^2/a_1^2 + m_2^2/a_2^2}. \tag{15}$$

The projection of the apparent velocity onto the baseline $\mathbf{r}$, evaluated from the temporal fluctuations ($\mathbf{r}$ = const), is found from the condition

$$\left.\frac{\partial \rho\,(\mathbf{r}, \tau)}{\partial \tau}\right|_{\mathbf{r}=\text{const}} = 0 \tag{16}$$

and for small velocity spread $|\mathbf{V_1} - \mathbf{V_2}| \ll {}^1\!/_2 |\mathbf{V_1} + \mathbf{V_2}|$ in expression (13) is equal to

$$V_a = \left(\frac{m_1^2}{a_1^2} V_1^2 + \frac{m_2^2}{a_2^2} V_2^2\right)\left[\frac{\mathbf{r}}{r}\left(\frac{m_1^2}{a_1^2} \mathbf{V_1} + \frac{m_2^2}{a_2^2} \mathbf{V_2}\right)\right]^{-1}. \tag{17}$$

Comparing (17) with the projection of the velocity $\langle \mathbf{V} \rangle$ onto the baseline $\mathbf{r}$, we see that

$$V_a > \left(\langle \mathbf{V} \rangle \frac{\mathbf{r}}{r}\right).$$

In the special case of one-dimensional drift $(\mathbf{V_1} \| \mathbf{V_2} \| \mathbf{r})$ and for $m_1 = m_2$, $a_1 = a_2$ we obtain from (15) and (17)

$$\langle V \rangle = (V_1 + V_2)/2; \quad \langle V \rangle \neq V_a = {}^1\!/_4 (V_1 - V_2)^2/\langle V \rangle. \tag{18}$$

We note that the two-velocity example is physically transparent and is realized in part under actual solar wind conditions. For example, it is clear at once from expression (13) that the correlation cross section of the BPS method does not describe the two-velocity drift process. We now consider an arbitrary distribution of velocities along the line of sight: $\varphi(\mathbf{V})$. We assume that the velocity SD is small, $\sigma = [\int (\mathbf{V} - \langle \mathbf{V} \rangle)^2 \varphi(\mathbf{V})d\mathbf{V}]^{1/2} \ll |\mathbf{V}| = |\int \mathbf{V} \varphi(\mathbf{V})d\mathbf{V}|$, because this is precisely the situation of greatest significance in regard to interplanetary scintillation studies. We also assume that the spatial correlation of the intensity fluctuations is isotropic and square-law with respect to the argument:

$$\rho^{(0)}(\mathbf{r}) = \rho^{(0)}(r^2). \tag{19}$$

Substituting (19) into (8) and taking the smallness of the velocity SD into consideration, on the basis of (14) and (16) we obtain for the velocities $\langle \mathbf{V} \rangle$ and V_a

$$\langle \mathbf{V} \rangle = \int \mathbf{V} \varphi(\mathbf{V}) \, d\mathbf{V}; \tag{20}$$

$$V_a = \langle V^2 \rangle \left(\frac{\mathbf{r}}{r} \langle \mathbf{V} \rangle\right)^{-1}, \quad \langle V^2 \rangle = \int V^2 \varphi(\mathbf{V}) \, d\mathbf{V}. \tag{21}$$

It follows from (20) and (21) that $V_a > \mathbf{r}/r \langle \mathbf{V} \rangle$. When the direction of the average velocity $\langle \mathbf{V} \rangle$ coincides with the direction of the baseline and the velocity SD is isotropic, $\sigma_x = \sigma_y = \sigma$, we obtain from (20) and (21)

$$V_a = \langle V \rangle + 2\sigma^2/\langle V \rangle. \tag{22}$$

For the one-dimensional velocity SD $\sigma_x = \sigma$, $\sigma_y = 0$ we have*

$$V_a = \langle V \rangle + \sigma^2/\langle V \rangle. \tag{23}$$

Thus, owing to the presence of velocity fine structure the apparent velocity of the IPS pattern always turns out to be greater than the projection of the average velocity of the inhomogeneities onto the baseline.

b. Scintillation Correlation Scales. We now inquire how the fine structure of the velocities affects the width (form) of the autocorrelation and cross-correlation functions. It is well known that the characteristic time scale of the correlation is specified as the recip-

---

*Relations (22) and (23) have already been used above.

rocal of the second moment of the temporal spectrum of the intensity fluctuations:

$$\tau_1^2 = -\rho(0,0)\left[\frac{\partial^2\rho(0,\tau)}{\partial\tau^2}\Big|_{\tau=0}\right]^{-1}, \quad \tau_2^2 = -\rho(\mathbf{r},\tau_0)\left[\frac{\partial^2\rho(\mathbf{r},\tau)}{\partial\tau^2}\Big|_{\tau=\tau_0}\right]^{-1}, \tag{24}$$

where $\tau_{1,2}$ denotes the characteristic times of the auto- and cross-correlations and τ_0 is the temporal position of the maximum of the cross-correlation function $\rho(\mathbf{r},\tau)$. The spatial correlation scales are specified as the products of the time scales $\tau_{1,2}$ and the apparent velocity V_a:

$$a_{1,2} = V_a\tau_{1,2}. \tag{25}$$

The correlation scales $\tau_{1,2}$ and $a_{1,2}$ have been investigated in [20] for two models of the medium: 1) a medium with two characteristic velocities; 2) a medium with a normal distribution of velocities with respect to the average velocity $\langle V\rangle$. It was shown that for the indicated velocity structure the cross-correlation function broadens relative to the autocorrelation function. The correlation scales in the two-velocity medium in this case are

$$\tau_1^2 = \frac{2a^2}{V_1^2+V_2^2}, \quad a_1^2 = \frac{2a^2(V_1^2+V_2^2)}{(V_1+V_2)^2}; \tag{26}$$

$$\tau_2^2 = \tau_1^2\left\{1 + \frac{r^2}{2a^2}\left[\left(1-\frac{V_1}{\langle V\rangle}\right)^2 + \left(1-\frac{V_2}{\langle V\rangle}\right)^2\right]\right\}, \quad a_2^2 = \tau_2^2\frac{(V_1^2+V_2^2)^2}{(V_1+V_2)^2}. \tag{27}$$

Analogously, for the model of a medium with a normal distribution of velocities [correlation function (11)] the scales $\tau_{1,2}$ and $a_{1,2}$ are*

$$\tau_1^2 = \frac{a^2}{\langle V\rangle^2+2\sigma^2}, \quad a_1^2 = \frac{a^2(\langle V\rangle^2+2\sigma^2)}{\langle V\rangle^2}; \tag{28}$$

$$\tau_2^2 = \tau_1^2 + \frac{r^2\sigma^2}{\langle V\rangle^4}, \quad a_2^2 = a_1^2 + \frac{r^2\sigma^2}{\langle V\rangle^2}. \tag{29}$$

A comparison of (26) with (27) and of (28) with (29) shows that the occurrence of fine structure in the velocities of the inhomogeneities along the line of sight tends to increase the cross-correlation scale in comparison with the autocorrelation scale. The amount of broadening is proportional to the variance (SD squared) of the velocities and the square of the length of the baseline r.

We now consider the dependence of the correlation scales on the velocity SD in the case of an arbitrary spatial correlation $\rho(\mathbf{r})$ of the scintillations and an arbitrary velocity distribution. We use the correlation function (7) in relations (24) and (25). Then for the temporal autocorrelation scale we obtain

$$\tau_1^2 = -\rho(0,0)\left[\frac{\partial^2\rho(\mathbf{r})}{\partial r^2}\Big|_{r=0}\langle V^2\rangle\right]^{-1} = \frac{a^2}{\langle V^2\rangle}, \tag{30}$$

where $a^2 = -\rho(0)\left[\frac{\partial^2\rho(\mathbf{r})}{\partial r^2}\Big|_{r=0}\right]^{-1}$ is the correlation space scale and $\langle V^2\rangle = \int V^2\varphi(\mathbf{V})d\mathbf{V}$. Substituting (30) into (25), we obtain for the space scale a_1:

$$a_1^2 = a^2\langle V^2\rangle\left(\frac{\mathbf{r}}{r}\langle\mathbf{V}\rangle\right)^{-2}. \tag{31}$$

---

*It is assumed in the derivation of relations (26)–(29) that the spatial correlation function of the intensity fluctuations is Gaussian.

As is evident from relation (31), $a_1 \neq a$, and so the scale a_1 is called the apparent correlation scale, by contrast with the correlation scale in the medium a. In the case of the average drift along the base line ($\langle \mathbf{V} \rangle \parallel \mathbf{r}$) relation (31) takes the form

$$a_1 = a \left(1 + \frac{2\sigma^2}{\langle V \rangle^2} \right)^{1/2}. \tag{32}$$

We deduce two important consequences from relations (30)-(32). First, it is evident from (30) that with fine structure present in the velocities ($\sigma \neq 0$) the time scale τ_1 turns out to be less than the correlation time scale for frozen-in motion with the velocity $\langle V \rangle$. Second, the apparent space scale a_1 (32) is greater than the scale a of the diffraction pattern. It is important to note in this connection that the apparent velocity V_a (21) is more sensitive than the apparent scale a_1 to variations of $\langle V \rangle$ and σ, as is evinced by a comparison of (21) and (32).

We now substitute the cross-correlation function $\rho(\mathbf{r}, \tau)$ (7) into relations (24) and (25). Up to small terms of order $\langle (\mathbf{r} - \mathbf{V}\tau_0)^2 \rangle / a^2 \sim \sigma^2 / \langle V \rangle^2$ the correlation scale τ_2 is equal to

$$\tau_2^2 = \tau_1^2 \left\{ 1 + \frac{r^2}{2} \left[\left| \frac{\partial^4 \rho(r)/\partial r^4}{\partial^2 \rho / \partial r^2} \right|_{r=0} \frac{\langle V^2 (\mathbf{r}/r - \mathbf{V}_\perp/V_a)^2 \rangle}{\langle V^2 \rangle} - \left| \frac{\partial^2 \rho(r)}{\partial r^2} \right|_{r=0} \left\langle \left(\frac{\mathbf{r}}{r} - \frac{\mathbf{V}}{V_a} \right)^2 \right\rangle \right] \right\}. \tag{33}$$

It follows from expression (33) that in the case of uniform motion, such that $\varphi(\mathbf{V}) = \delta(\mathbf{V} - \mathbf{V}_0)$, the cross-correlation scale τ_2 is equal to the autocorrelation scale τ_1. In models of the medium with two velocities and with a normal distribution of velocities relation (33) goes over to the respective expressions (27) and (29). The quantity $|(\partial^4 \rho / \partial r^4)/(\partial^2 \rho / \partial r^2)|_{r=0}$ in (33) is equal to q_4^4/q_2^2, where q_4 and q_2 are the fourth and second moments of the spatial scintillation spectrum. For the incident spatial spectrum always $q_4^4/q_2^2 > |\partial^2 \rho / \partial r^2|_{r=0} = q_2^2$, and so the cross-correlation scale τ_2 is always greater than the autocorrelation scale τ_1. The broadening of the cross-correlation function is proportional to the baseline squared r^2 and to the variance of the velocity σ^2.

Thus, the presence of velocity fine structure in the medium always changes the correlation scales relative to the case of uniform motion; the autocorrelation time scale τ_1 becomes less than the scale $a/\langle V \rangle$ for uniform motion; the apparent space scale a_1 (32) is greater than the scale a of the diffraction pattern; and the cross-correlation time and space scales τ_2 and a_2 are greater than the corresponding scales τ_1 and a_1 of the autocorrelation. The indicated scale changes are more pronounced the greater the velocity SD. The broadening of the cross-correlation function (variation of its shape) in comparison with the autocorrelation function for a given velocity SD increases with the baseline $|\mathbf{r}|$ according to a square law.

c. Cross-correlation Skewness. The cross-correlation broadening effect discussed above ($\tau_2 > \tau_1$), expression (33), is proportional to the standard deviation of the velocities of the inhomogeneities. In relation (33), which describes this effect, the cross-correlation function $\rho(\mathbf{r}, \tau)$ is expanded into a series in the parameter $\langle (\mathbf{r} - \mathbf{V}\tau_0)^2 \rangle / a^2$.

We now examine another effect: cross-correlation skewness, which is also related to the fine structure of the velocities of the inhomogeneities but has the next-higher order of smallness with respect to the parameter $\langle (\mathbf{r} - \mathbf{V}\tau_0)^2 \rangle / a^2$. The skewness of the cross-correlation function $\rho(\mathbf{r}, \tau)$ is conveniently characterized by the skewness parameter S [35-37]:

$$S = \frac{\rho(\mathbf{r}, \tau_0 + t) - \rho(\mathbf{r}, \tau_0 - t)}{\rho(\mathbf{r}, \tau_0 + t) + \rho(\mathbf{r}, \tau_0 - t)}, \tag{34}$$

where τ_0 is the position of the maximum of the function $\rho(\mathbf{r}, \tau)$, and t is the shift relative to the maximum, t > 0. Comparing the skewness parameter S (34) with the parameters of the correlation scales $\tau_{1,2}$ (24), we note that the shift t relative to the maximum enters explicitly into the definition of the skewness parameter S, by contrast with the scales $\tau_{1,2}$, which characterize the function $\rho(\mathbf{r}, \tau)$ in the large. In this sense S is a differential parameter relative to the function $\rho(\mathbf{r}, \tau)$. A positive value of the skewness parameter S (for $\tau_0 > 0$) signifies that $\rho(\mathbf{r}, \tau)$ decays more rapidly to the left of the maximum (in the direction $\tau \to 0$) than to the right (in the direction $\tau \to \infty$).

We first discuss the two-velocity model. Substituting the function $\rho(\mathbf{r}, \tau)$ (13) into relation (34) for S, in the one-dimensional case ($\mathbf{r} \parallel \mathbf{V}_1 \parallel \mathbf{V}_2$) [34] we obtain

$$S = \frac{2rt^3 V_1 V_2}{a_1^2 a_2^2} \frac{\gamma(\alpha\beta^2 - 1)\Delta V}{(\gamma + 1)(\alpha\beta^2\gamma + 1)}, \tag{35}$$

where $\Delta V = V_1 - V_2$, $\beta = V_1/V_2$, $\alpha = a_2^2/a_1^2$, $\gamma = m_1^2/m_2^2$. It is evident from (35) that the sign of S depends on the ratio between the velocities V_1 and V_2 as well as between the inhomogeneity scales a_1 and a_2; for $V_1 > V_2$ we have $S > 0$ if $a_2/V_2 > a_1/V_1$ and $S < 0$ if $a_2/V_2 < a_1/V_1$.

Thus, in the two-velocity model of the medium the skewness parameter S is positive if the correlation time scale a_1/V_1 for uniform motion with the greater velocity is less than the correlation time scale a_2/V_2 for the motion with the smaller velocity; otherwise S is negative.

We now consider an arbitrary velocity distribution $\varphi(\mathbf{V})$. We expand the unperturbed correlation function $\rho^0(\mathbf{r} - \mathbf{V}\tau, 0)$ in relation (7) into a power series in $|\mathbf{r} - \mathbf{V}\tau|/a$ and retain the first three terms of the expansion:

$$\rho^{(0)}(\mathbf{r}) = 1 - \frac{r^2}{2a^2} + \frac{r^4}{8b^4}, \tag{36}$$

where $a^{-2} = -\partial^2\rho(\mathbf{r})/\partial r^2|_{r=0}$ and $b^{-4} = \frac{1}{3}\partial^4\rho(\mathbf{r})/\partial r^4|_{r=0}$; a and b are the correlation scales. In the weak scintillation regime (8) this expansion corresponds to calculation of the skewness in the interval of τ close to the maximum of the function $\rho(\mathbf{r}, \tau)$ at $\tau = \tau_0$. For simplicity we keep to the one-dimensional case, where the velocities of the inhomogeneities are directed along the baseline: $\mathbf{r}(x, 0)$; $\varphi(\mathbf{V}) = \varphi(V_x)\delta(V_y)$. We also assume that correlation scales a and b are constant and do not depend on the velocity V. Then, inserting (36) into (8) and taking account of the maximum condition $\partial\rho(\mathbf{r}, \tau)/\partial\tau|_{\tau_0} = 0$, we obtain

$$S = \frac{r\sigma^2 \langle V \rangle^3 t^3}{(\langle V \rangle^2 + \sigma^2)b^4}\delta, \tag{37}$$

$$\delta = 1 + \frac{1}{2}\gamma_1\frac{\sigma}{\langle V \rangle}\left(3 - \frac{\sigma^2}{\langle V \rangle^2}\right) + \frac{1}{2}\gamma_2\frac{\sigma^2}{\langle V \rangle^2}, \tag{38}$$

where $\gamma_1 = (1/\sigma^3)\int(V - \langle V \rangle)^3\varphi(V)dV$ is the momental skewness and $\gamma_2 = [(1/\sigma^4)\int(V - \langle V \rangle)^4 \times \varphi(V)dV - 3]$ is the kurtosis in the distribution function $\varphi(V)$. Expressions (37) and (38) establish the dependence of the skewness parameter S on the parameters of the velocity distribution function $\varphi(V)$. It is evident from (37) that

$$\text{sgn } S = \text{sgn } \delta \tag{39}$$

and, hence, knowing the moments of the velocity distribution $\varphi(V)$, we can at once determine the sign of the cross-correlation skewness S. For example, if $\gamma_1 > 0$ and $\gamma_2 > 0$, implying that the distribution $\varphi(V)$ is steeper than normal and decays more rapidly to the left of the maximum (in the direction $V \to 0$) than to the right ($V \to \infty$), the skewness parameter S is positive ($S > 0$), and the correlation function $\rho(\mathbf{r}, \tau)$ decays more rapidly to the left of the maximum (in the direction $\tau \to 0$) than to the right ($\tau \to \infty$). If the coefficients γ_1 and γ_2 are negative, i.e., if the function $\varphi(V)$ has the shape of a "triangle" and decays more rapidly in the direction of large V, and if $\gamma_{1,2}$ are not too great ($|\gamma_1| \sim |\gamma_2| \lesssim 1$), then for a small velocity SD ($\sigma \ll \langle V \rangle$) the parameter S is positive ($S > 0$). The cross-correlation skewness S becomes negative for rather strong asymmetry in the velocity distribution:

$$\gamma_1 < -{}^2/_3 \langle V \rangle / \sigma. \tag{40}$$

Negative values can also be imparted to S by negative values of the kurtosis ($\gamma_2 < 0$), in which case the distribution $\varphi(V)$ is flatter than normal,

$$\gamma_2 < -2 \langle V \rangle^2 / \sigma^2, \tag{41}$$

but since $\gamma_2 > -3$, for condition (41) to be satisfied it is necessary that the velocity SD not be too small: $\sigma \gtrsim \langle V \rangle$.

As an example we consider the problem in which it is required to determine the parameter S [the function $\varphi(V)$ is known, and we wish to determine the sign of S]; we inquire how the skewness is affected by allowance for the extent of the medium in the model of a spherically symmetrical solar wind with radial velocity V_0. The velocity distribution function in this solar-wind model has been derived earlier [37] and has the form

$$\varphi(V) = \begin{cases} \frac{4}{\pi} \left(\frac{V}{V_0} \right)^2 (V_0^2 - V^2)^{-1/2}, & V < V_0, \\ 0 & V > V_0, \end{cases} \tag{42}$$

where V_0 is the velocity of the wind. From (42) we obtain the following for the parameters of $\varphi(V)$ in (38):

$$\sigma/\langle V \rangle \simeq 0.2; \quad \gamma_1 \simeq -1. \tag{43}$$

Hence it follows that condition (40), for which S < 0, is not satisfied in the case (43), i.e., the skewness parameter is positive (S > 0). Thus, allowance for the extent of the medium yields a positive value of the cross-correlation skewness parameter.

It is essential to note that a negative cross-correlation skewness can be created not only by strong asymmetry in the velocity distribution $\varphi(V)$, but also by the existence of a growing dependence of the correlation scales $a(V)$ and $b(V)$ on the velocity of the inhomogeneities V [it was assumed in the derivation of (37) that the scales a and b are constant]. If the scales a and b decay with increasing V or do not grow more rapidly than V^1, the result (37) remains in effect, relating the skewness S to the parameters of $\varphi(V)$. But if $a \propto b \propto V^\alpha$, $\alpha = 1$, then the skewness S vanishes, S = 0. Finally, for a stronger dependence of the scales a and b on the velocity ($a \propto b \propto V^\alpha$, $\alpha > 1$), even for small velocity SD ($\sigma \ll \langle V \rangle$), the skewness becomes negative (S < 0).

Thus, a spread of velocities of the inhomogeneities induces a cross-correlation skewness S that is positive for weak spread ($\sigma \ll \langle V \rangle$) and has a value proportional to the baseline r and variance of the velocities σ^2. It is important to note that a correlation cross section $\rho(\mathbf{r}, \tau)$ of the form (3), as is customarily used in the BPS method, does not allow for the skewness effect, because the points of intersection of the elliptical constant-correlation contour with the line x = const are symmetrical about the line $x = V_a \tau$ (see Fig. 1; this fact has already been noted in Section 1.1). Cross-correlation skewness occurs only in relation to the temporal cross-correlation $\rho(\mathbf{r}, \tau)|_{\mathbf{r} = \text{const}}$. The cross-correlation cross section $\rho(\mathbf{r}, \tau)|_{\mathbf{r} = \text{const}}$ (spatial cross-correlation function) turns out to be symmetrical for an arbitrary velocity distribution $\varphi(\mathbf{V})$.

The cross-correlation skewness effect can be utilized to investigate the velocity distribution function $\varphi(V)$ of the inhomogeneities on the assumption of a wind-type scintillation pattern.

We now interject a few remarks concerning the practical computation of the skewness parameter S. First, the parameter S (37) is obtained on the basis of expression (8), in which

$$p^{(0)}(\mathbf{r}) = \int M(\mathbf{q}) \exp(i\mathbf{q}\mathbf{r}) \, d\mathbf{q}, \qquad (44)$$

and $M(\mathbf{q})$ is the spatial spectrum of the intensity fluctuations. If the moments of all orders of the function $\varphi(V)$ $(V_n = [\int V^n \varphi(V) dV]^{1/n})$ and the spectrum $M(\mathbf{q})$ $(q_n = [\int q^n M(\mathbf{q}) dq]^{1/n})$ are finite, then the expansion (36) is valid near the maximum of $\rho(\mathbf{r}, \tau)$. The condition for applicability of the expansion (36) takes the form

$$t \ll \sqrt{5} \, \frac{q_4^2 V_4^2}{q_6^3 V_6^3} \sim \frac{a}{V_2} \qquad \text{for} \qquad q_6^3 \sim q_2 q_4^2, \, V_6^3 \sim V_2 V_4^2. \qquad (45)$$

This means that the deviations t from the maximum of $\rho(\mathbf{r}, \tau)$ must be small in comparison with the characteristic time scale of the autocorrelation $\rho(0, \tau)$. For such deviations we have already computed the skewness S (37), which is a function of the shift t $(S \sim t^3)$, i.e., the value of S increases with t. Consequently, from the standpoint of reliability of computation of S it is favorable to choose the values of t as large as possible.

It is known [38] that skewness of the function $\rho(\mathbf{r}, \tau)$ takes place not only near the maximum τ_0, but also in the "tails" of the function $\rho(\mathbf{r}, \tau)$, i.e., for values of $t \gtrsim a/V_2$. The question therefore arises, for what values of the shift t is it necessary to measure the parameter S for the real function $\rho(\mathbf{r}, \tau)$? Let us consider the skewness S for three intervals of the function $\rho(\mathbf{r}, \tau)$: 1) in the interval immediately adjacent to the maximum, $t \ll a/V_2$; 2) on the "slope," where $t \lesssim a/V_2$; 3) in the "tails," where $t > a/V_2$. For $t \ll a/V_2$ the expansion (36) is valid with good accuracy, while the skewness $S \sim t^3$ and depends only on the fourth moment q_4 of the spectrum $M(\mathbf{q})$, i.e., in this interval of t the skewness S is independent of the shape of the spatial spectrum $M(\mathbf{q})$. For small t, however, S is small and therefore difficult to detect. On the "slope" of the function $\rho(\mathbf{r}, \tau)$, i.e., in the interval $t \lesssim a/V_2$ [for values of t smaller than or of the same order as the width of $\rho(\mathbf{r}, \tau)$ at the e^{-1} level], the expansion (36) can again be used, but with somewhat lower accuracy. It may be approximately assumed that in this interval of t the skewness S exhibits little sensitivity to the detailed shape of the spectrum $M(\mathbf{q})$ and is determined mainly by the fourth moment q_4. In the "tails" of $\rho(\mathbf{r}, \tau)$, i.e., for $t > a/V_2$, the skewness can be appreciable. However, the "tails" are relatively useless for determination of the skewness, because the skewness in the "tails," unlike the same on the "slope" and in the vicinity of the maximum of $\rho(\mathbf{r}, \tau)$, depends not only on the fine structure of the velocity $\varphi(\mathbf{V})$, but also on the shape of the spectrum $M(\mathbf{q})$; in addition, the function $\rho(\mathbf{r}, \tau)$ itself is not sufficiently reliable in the "tails" due to the finiteness of the observation time, and so in this interval the skowness S is determined with large error.

Consequently, under weak-scintillation conditions the optimal shifts t for the calculation of S lie in the interval $t \lesssim a/V_2 [\rho(\mathbf{r}_0, \tau_0 \pm t) > (1/e)\rho(\mathbf{r}, \tau_0)]$, where for example, t can be selected at the half-peak level: $\rho(\mathbf{r}, \tau_0 + t) = {}^1/_2 \rho(\mathbf{r}, \tau_0)$. Relation (37) must be regarded as approximate in this case.

Second, Eq. (34) is not the unique definition of skewness. Other relations have been proposed in the literature for the practical definition of the parameter S. Golley and Denisson [39], for example, compute the skewness as follows:

$$S_1 = \frac{\tan \alpha_1 - \tan \alpha_2}{\tan \alpha_1 + \tan \alpha_2}, \qquad (46)$$

where α_1 and α_2 are the angles indicated in Fig. 2.

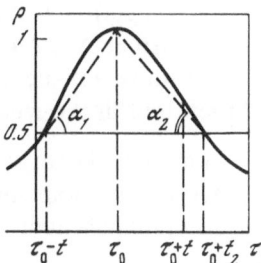

Fig. 2. Determination of the skewness of the cross-correlation function $\rho(\mathbf{r}, \tau)$.

This technique is based on approximation of the upper part of the function $\rho(\mathbf{r}, \tau)$ [in the interval $\rho(\mathbf{r}, \tau) > \frac{1}{2}$] by a triangle. McGee [40] defines the skewness parameter as

$$S_2 = (t_2 - t_1)/(t_2 + t_1). \tag{47}$$

A comparison of the various ways of defining S, (34), (46), and (47), shows that in the case of a small skewness ($S \ll 1$), determined at the half-peak level, all three definitions approximately coincide: $S = S_1 = S_2$.

1.4. Saturated Scintillations

The discussion thus far concerning the form of the cross-correlation function of the intensity flucutuations referred to the weak-scintillation regime, where the expression (7) for the cross-correlation function $\rho(\mathbf{r}, \tau)$ is valid. The influence of velocity spread on the temporal spectra of saturated scintillations has been investigated in [41]. We now investigate how spread of the inhomogeneities affects the form of the cross-correlation function of the intensity fluctuations in the saturated-scintillation regime, where expression (8) is now valid for $\rho(\mathbf{r}, \tau)$.

We first calculate the pattern drift velocities $\langle \mathbf{V} \rangle$ and V_a. To do so we substitute the cross-correlation function $\rho(\mathbf{r}, \tau)$ (8) into the maximum conditions (14) and (16). It is readily verified that the position of the maximum of $\rho(\mathbf{r}, \tau)$ coincides with the position of the maximum of the correlation function $\rho_\Phi(\mathbf{r}, \tau)$ of the phase fluctuations, and so the drift velocity $\langle \mathbf{V} \rangle$ and apparent velocity V_a of the IPS pattern, as in the weak-scintillation case, are given by expressions (20) and (22).

The correlation scales $\tau_{1,2}$ (24) and the apparent space scales $a_{1,2}$ of the autocorrelation and cross-correlation can be calculated analogously. For the calculation of these scales we make use of the following circumstance. Inasmuch as the rms phase-angle lead Φ_0 in relation (8) is much greater than unity, the correlation space scale of the intensity fluctuations is equal to a/Φ_0 and is much less than the correlation scale a of the phase fluctuations. Consequently, in calculating the correlation scales $\tau_{1,2}$ we can use the expansions of the structure function $D_\Phi(\mathbf{r}, \tau)$ near the minimum for $\tau = \tau_0$ with respect to $(\mathbf{r} - \mathbf{V}\tau)^2/a^2$ and restrict the expansion to first term. Then the correlation scales $\tau_{1,2}$ are

$$\tau_1^2 = -\frac{1}{\Phi_0^2}\left[\frac{\partial^2 \rho_\Phi(0, \tau)}{\partial \tau^2}\bigg|_{\tau=0}\right]^{-1},$$
$$\tau_2^2 = -\frac{1}{\Phi_0^2}\left[\frac{\partial^2 \rho_\Phi(\mathbf{r}, \tau)}{\partial \tau^2}\bigg|_{\tau=\tau_0}\right]^{-1}. \tag{48}$$

A comparison of (48) and (24) shows that the time scales $\tau_{1,2}$ in the case of saturated scintillations differ from the weak-scintillation correlation scales $\tau_{1,2}$ by the additional factor $1/\Phi_0$. The same is true of the apparent space scales $a_{1,2}$ (25). Consequently, a reduction of the time scale and an increase of the apparent space scale of correlation relative to uniform motion, as well as broadening of the cross-correlation function $\rho(\mathbf{r}, \tau)$ relative to the autocorrelation function $\rho(0, \tau)$ (see Section 1.3b) take place in the saturated-scintillation regime as well. The absolute scale variations $\Delta\tau$, Δa in the saturated-scintillation regime will be smaller by a factor $1/\Phi_0$ than in the weak-scintillation case; the relative scale variations $\Delta\tau/\tau$ and $\Delta a/a$ turn out to be identical in the weak and saturated scintillation regimes.

We now consider the cross-correlation skewness S (see Section 1.3c) for saturated scintillations. We use the expansion (36) in expression (8) for $\rho(\mathbf{r}, \tau)$ and (34) for S. As a result, we obtain for the skewness

$$S^{\text{sat}} = \tanh(2\Phi_0^2 S^{\text{wk}}),\tag{49}$$

where S^{wk} is given by relation (37). It is instructive to compare the cross-correlation skewness S in the saturated and weak scintillation regimes. In the former the correlation time scale of the intensity fluctuations is $a/(\langle V \rangle \Phi_0)$, so that, substituting $t \simeq a/(\langle V \rangle \Phi_0)$ into (49) for $a \simeq b$, we obtain the estimate for S

$$S = \frac{2}{\Phi_0} \frac{r}{a} \frac{\sigma^2}{\langle V \rangle^2 + \sigma^2} \delta; \qquad \Phi_0^2 \gg 1.\tag{50}$$

Analogously, in the case of weak scintillations, where the correlation time scale is $a/\langle V \rangle$, from (37) we obtain for $t \simeq a/\langle V \rangle$

$$S = \frac{r}{a} \frac{\sigma^2}{\langle V \rangle^2 + \sigma^2} \delta; \qquad \Phi_0^2 \ll 1.\tag{51}$$

A comparison of (50) and (51) brings us to the conclusion that in the weak-scintillation regime the cross-correlation skewness is $\Phi_0/2$ times as strong as in the saturated-scintillation case. Equation (49) for S^{sat} is reasonably accurate on the "slope" of the function $\rho(\mathbf{r}, \tau)$, as opposed to Eq. (37) for S^{wk}, which is approximate on the "slope" of $\rho(\mathbf{r}, \tau)$.

Thus, a spread of velocities of the inhomogeneities produces the following correlation effects: (1) The apparent velocity of the scintillation pattern, determined at the maximum of the temporal correlation $\rho(\mathbf{r}, \tau)$, is greater than the average velocity of the inhomogeneities; (2) the autocorrelation time scale τ_1 decreases by comparison with the autocorrelation scale for uniform motion; (3) the apparent space scale a_1 of the scintillations is greater than the space scale a of the diffraction pattern; (4) the cross-correlation function $\rho(\mathbf{r}, \tau)$ of the intensity fluctuations broadens by comparison with the autocorrelation function $\rho(0, \tau)$; (5) the cross-correlation function $\rho(\mathbf{r}, \tau)$ acquires skewness S. The first three effects occur in the BPS method, which disregards the last two. The indicated effects are more pronounced the greater the velocity spread of the inhomogeneities.

The results of analysis of the form and parameters of the temporal auto- and cross-correlation functions provide a means for the disclosure of fine structure of the velocities in the scattering medium and for determination of its characteristics. Thus, skewing and broadening of the cross-correlation function $\rho(\mathbf{r}, \tau)$ by comparison with the autocorrelation function $\rho(0, \tau)$ always occur when a spread of velocities of the inhomogeneities is present in the medium, so that cross-correlation skewness and broadening serve as tests for the detection of velocity fine structure. Also, investigation of the parameters of $\rho(0, \tau)$ and $\rho(\mathbf{r}, \tau)$ enables us to solve the cross-correlation analysis problem, namely to determine the scintillation space scale a, the average velocity of the inhomogeneities $\langle V \rangle$, and the SD of the velocities σ. Treating relations (21) for V_a, (26) for τ_1 or a_1, and (37) for S with $\delta \simeq 1$ as a system of equations, we can find a, $\langle V \rangle$, and σ, i.e., solve the cross-correlation analysis problem. It is equally important to determine the parameters γ_1 and γ_2 of the velocity distribution function. Analysis of the cross-correlation skewness should make it possible in principle to determine $\gamma_{1,2}$, but in the case $\sigma/\langle V \rangle \ll 1$ it is possible to do so only for a fairly large observation period T.

The method described here for estimating the drift parameters of the scintillation pattern has a number of advantages over other methods. Above all, this method is far simpler than the others and does not require nearly the volume of computations as the BPS and LE methods. Furthermore, in determining the drift parameters from the form of the cross-correlation functions, by contrast with the BPS method, it is not necessary to postulate ellipticity on the part of the constant-correlation contours, and so this approach has a sizable domain of applicability. The method of intersections discussed in Section 1.1b, like the method proposed here, is based on measurement of the individual parameters of the correlation functions. However, as mentioned, in the interpretation of the experimental data the method of intersections, like the BPS and LE methods, encounters serious difficulties, which do not occur in the given method.

2. Dispersion (or Cross-spectral) Analysis of the Velocity of the Diffraction Pattern

Another approach that can be used to study the velocity of the solar wind is based on cross-spectral analysis of the velocity of the diffraction pattern. The idea behind this approach was advanced by Lotova and Chashei [35, 42, 43]. Only comparatively recently has dispersion analysis begun to be discussed, in contradistinction with the well-developed cross-correlation method.

The intensity fluctuations $\delta I(\mathbf{r}, t)$ of a radio source can be represented by a superposition of Fourier components of the temporal spectrum $P(\omega)$, each of which has its own velocity $V(\omega)$, or $V(\mathbf{q})$, for the spatial spectrum $M(\mathbf{q})$. The dispersion dependence $V(\omega)[V(\mathbf{q})]$ is obtained on the basis of an analysis of the temporal (spatial) cross-correlation spectra $P_c(\omega, \mathbf{r})[M_c(\mathbf{q}, t)]$. The spectral composition $V(\omega)$ of the IPS-pattern drift velocity contains information about the velocity distribution in the scattering medium.

The dispersion analysis problem entails the determination of the average velocity of the inhomogeneities $\langle V \rangle$ and the velocities of the harmonics of the temporal spectrum $V(\omega)$ [or spatial spectrum $\mathbf{V}(\mathbf{q})$] of the intensity fluctuations and the determination of the spatial spectrum $M(\mathbf{q})$ of the intensity fluctuations. The spectrum $M(\mathbf{q})$ is related to the spectrum of fluctuations of the density of electrons $F_N^2(\mathbf{q})$ of the medium, and in the case of weak scintillations [31]

$$M(\mathbf{q}) = 4 \sin^2 \left(q^2 \frac{z}{2k} \right) F_N^2(\mathbf{q}). \tag{52}$$

2.1. Spectral Method of Spatial Dispersion

The spectral approach to analysis of the drift velocity of the diffraction pattern was first proposed by Briggs and is based on an investigation of the dependence of the velocity on the spatial frequency. This method was elaborated and utilized by Briggs [44] to analyze motions in the ionosphere. The method essentially entails investigation of the velocity of the diffraction pattern according to measurements of the intensity fluctuations at many stations at two different times. "Photographs" of the pattern at two times t_1 and $t_1 + t_0$ [$\delta I(\mathbf{r}, t_1)$ and $\delta I(\mathbf{r}, t_1 + t_0)$] make it possible to determine the complex-valued spectra of the intensity fluctuations:

$$F_I(\mathbf{q}, t_{1,2}) = |F_{1,2}| \exp(i\Phi_{1,2}) = \int \delta I(\mathbf{r}, t_{1,2}) \exp(i\mathbf{q}\mathbf{r}) d\mathbf{r}, \tag{53}$$

where $|F_{1,2}|$ and $\Phi_{1,2}$ are the modulus and phase of the spectral expansion of the intensity fluctuations and $\mathbf{q}$ is the two-dimensional spatial frequency. The composition $\langle F_I(\mathbf{q}, t_1) F_I^*(\mathbf{q}, t_2) \rangle$ represents the Fourier transform of the spatial cross-correlation function $\rho(\mathbf{r}, t_0)$:

$$
\begin{aligned}
M(\mathbf{q}, t_0) &= \langle F_I(\mathbf{q}, t_1) F_I^*(\mathbf{q}, t_2) \rangle = \\
&= |F_1||F_2| \exp[i\Delta\Phi(\mathbf{q}, t_0)] = \frac{1}{(2\pi)^2} \int \rho(\mathbf{r}, t_0) \exp(i\mathbf{q}\mathbf{r}) d\mathbf{r}, \\
\Delta\Phi(\mathbf{q}, t_0) &= \Phi(\mathbf{q}, t_1) - \Phi(\mathbf{q}, t_2),
\end{aligned}
\tag{54}
$$

where $M(\mathbf{q}, t_0 = 0)$ is the sought-after spatial spectrum of the intensity fluctuations and $\Gamma(\mathbf{q}) = F_1 F_2 / M$ is the spatial coherence function. The phase difference $\Delta\Phi(\mathbf{q}, t_0)$ characterizes the motion of the $\mathbf{q}$-component of the spatial spectrum $M(\mathbf{q})$. The quantity $\Delta\Phi(\mathbf{q}, t_0)$ provides a means for determining the average drift velocity and spatial dispersion of the velocities $\mathbf{V}(\mathbf{q})$. Thus, the expansion (53) corresponds to representation of the fluctuations $\delta I(\mathbf{r}, t_{1,2})$ by a set of plane waves with a two-dimensional wave vector $\mathbf{q}$ and phase angles

$$\Psi(\mathbf{q}, t_{1,2}) = \mathbf{q}\mathbf{r} - \Phi(\mathbf{q}, t_{1,2}). \tag{55}$$

The equation for the wave front is determined from the constant-phase condition $\mathbf{q}\mathbf{r} - \Phi(\mathbf{q}, t) = $ const, whence we infer that the distance $\Delta\mathbf{r}$ traveled by the wave $\mathbf{q}$ in time Δt satisfies the equation

$$\mathbf{q}\Delta\mathbf{r} - \Delta\Phi(\mathbf{q}, \Delta t) = 0. \tag{56}$$

From Eq. (56) for fixed harmonics $\mathbf{q}$ with respect to the phase $\Delta\Phi(\mathbf{q}, \Delta t)$ we obtain the velocity $\mathbf{V}(\mathbf{q})$:

$$\mathbf{q}\mathbf{V}(\mathbf{q}) = \Delta\Phi(\mathbf{q}, \Delta t)/\Delta t.$$

For the case in which the velocity $\mathbf{V}(\mathbf{q})$ has the same direction for all $\mathbf{q}$ we obtain from (57)

$$V(q_x) = \frac{\Delta\Phi(q_x, \Delta t)}{q_x \Delta t}. \tag{58}$$

Thus, from measurements of the intensity fluctuations at many stations at two different times the cross-spectral method can be used not only to determine the average velocity of the inhomogeneities (as in the method of cross-correlation analysis; see Section 1), but also to obtain the spectral composition of the diffraction-pattern velocity. The function $V(q_x)$ has a real physical significance, being the phase velocity of the waves forming the inhomogeneous structure, i.e., it characterizes the time variation of the inhomogeneities. When the motion of the inhomogeneities is not dispersive, $V(q_x) = V_0 = $ const. However, the interpretation of the dispersion function $V(q_x)$ as the phase velocity of waves is admissible only when the velocities of the inhomogeneities do not have an angular (directional) spectrum. On the other hand, if the wave structures $F_N^2(\mathbf{q})$ move in different directions, i.e., if there is an angular spectrum of waves, the interpretation of the function $V(q_x)$ becomes ambiguous in the method of cross-spectral analysis. For example, Briggs and Golley [45] have concluded that the observed dispersive motion in the pattern of ionospheric scintillations are related to unequal velocities of motion of the scattering structure at different heights, rather than to actual dispersion wave motions.

However, the method of Briggs is unrealistic in practical radio-astronomic measurements, because to set up the great many stations needed to obtain the instantaneous "photographs" of the scintillation pattern calls for enormous material expenditures.

2.2. Spectral Method of Temporal Dispersion; Analysis of

Fine Structure of the Solar Wind Velocity

In radio-astronomic observations we are concerned with temporal measurements at two or three stations. We therefore consider the temporal dispersion. Suppose that temporal intensity fluctuations $\delta I_{1,2}(t)$ are recorded simultaneously at two stations with baseline $\mathbf{r}$. We expand the temporal fluctuations $\delta I_{1,2}(t)$ into the time Fourier spectrum

$$\delta I_{1,2}(t) = \int P_{1,2}(\omega) \exp(-i\omega t) d\omega, \tag{59}$$

in which

$$P_{1,2}(\omega) = |P_{1,2}(\omega)| \exp[i\Phi_{1,2}(\omega)] = \frac{1}{2\pi}\int \delta I_{1,2}(t)\exp(i\omega t)\,dt. \tag{60}$$

According to the convolution theorem the temporal cross-spectrum of the intensity fluctuations is

$$\langle P_1(\omega) P_2^*(\omega')\rangle = P_c(\omega)\,\delta(\omega - \omega') = |P_c(\omega)|\exp[i(\Phi_1(\omega) - \Phi_2(\omega'))]\,\delta(\omega - \omega'), \tag{61}$$

where $P_c(\omega)$ is the spectrum of the cross-correlation function of the intensity fluctuations $\rho(\mathbf{r}, \tau)$:

$$P_c(\omega) = \frac{1}{2\pi}\int \rho(\mathbf{r}, \tau)\exp(i\omega\tau)\,d\tau. \tag{62}$$

The representation (59) of the intensity fluctuations is a superposition of simple harmonic oscillations with frequencies ω. If we assume that these oscillations are associated with plane waves, the equation for the wave front takes the form

$$\omega t - \Delta\Phi(\omega, \mathbf{r}) = \text{const.} \tag{63}$$

Unlike (55), the dependence of the phase angle on the space variable $\mathbf{r}$ in relation (63) is contained in $\Delta\Phi(\omega, \mathbf{r})$. From the wave-front equation (63) we can determine the time Δt required for the wave with frequency ω to travel the distance $\mathbf{r}$ between stations:

$$\Delta t = \Delta\Phi(\omega, \mathbf{r})/\omega, \tag{64}$$

and from this result we find for the harmonic velocities $V(\omega)$:

$$V(\omega) = r\cos\theta/\Delta t = \omega r\cos\theta/\Delta\Phi(\omega, \mathbf{r}), \tag{65}$$

where θ is the angle between the velocity $\mathbf{V}$ and the baseline $\mathbf{r}$: $\theta = \hat{\mathbf{V}}\mathbf{r}$.

In the case of uniform motion, where only one inhomogeneity velocity is present in the medium, the dispersion function [frequency dependence of the harmonic velocity $V(\omega)$] is absent, $V(\omega) = $ const, and the phase $\Delta\Psi(\omega)$ of the cross-spectrum is proportional to the frequency ω [$\Delta\Phi(\omega) \sim \omega$] irrespective of the form of the spatial spectrum of the intensity fluctuations $M(\mathbf{q})$ (52). It is reasonable to expect that the fine structure of the velocities in the solar wind will effect a change in the function $\Delta\Phi(\omega)$ [or $V(\omega)$] [42, 43]. The dispersion analysis of the velocity fine structure in the solar wind is based on an investigation of the function $V(\omega)$ [35]. For the interpretation of the observed function $V(\omega)$ it is necessary to analyze the function $V(\omega)$ for various models of the scattering medium [42, 43]. We assume initially that the IPS pattern has a wind nature and the local frozen-in character of the inhomogeneities is not violated (the inhomogeneities do not vary with time). Within the framework of this assumption we analyze the behavior of the function $V(\omega)$ for various models of the medium: (a) a two-velocity medium; (b) a spherically symmetric solar wind; (c) an arbitrary velocity distribution. We finally take into account: (d) time variation of the inhomogeneities.

a. Two-Velocity Medium. In the model of a medium with two characteristic velocities the cross-correlation function $\rho(\mathbf{r}, \tau)$ has the form (13). Consequently, the cross-spectrum $P_c(\omega)$ represents the sum of two components, each of which corresponds to the cross-spectrum of uniform motion with the appropriate velocity $\mathbf{V}_1$ or $\mathbf{V}_2$:

$$P_c(\omega) = \int [M_1(\mathbf{q})\,\delta(\omega - \mathbf{q}\mathbf{V}_1) + M_2(\mathbf{q})\,\delta(\omega - \mathbf{q}\mathbf{V}_2)]\exp(i\mathbf{q}\mathbf{r})\,d\mathbf{q}, \tag{66}$$

where $M_1(\mathbf{q})$ and $M_2(\mathbf{q})$ are the spatial spectra of the intensity fluctuations for the velocities $\mathbf{V}_1$ and $\mathbf{V}_2$. We consider the one-dimensional case, in which $\mathbf{V}_1 \parallel \mathbf{V}_2 \parallel \mathbf{r}$. Now we have on the basis of (66)

$$P_c(\omega) = P_1(\omega) \exp(i\omega r/V_1) + P_2(\omega) \exp(i\omega r/V_2), \qquad (67)$$

where in accordance with (52)

$$P_{1,2}(\omega) = A_{1,2} \int \sin^2\left(q^2 \frac{z}{2k}\right) \cdot F_{N_{1,2}}^2(\mathbf{q})\,\delta(\omega - q_x V_{1,2})\,d\mathbf{q}; \qquad (68)$$

$A_{1,2} = $ const, and $F_{N_{1,2}}^2(\mathbf{q})$ denotes the spectra of fluctuations of the plasma density for the respective velocities V_1 and V_2. The phase $\Delta\Phi(\omega)$ of the cross-spectrum is given by the relation

$$\Delta\Phi(\omega) = \arctan \frac{\operatorname{Im} P_c(\omega)}{\operatorname{Re} P_c(\omega)}. \qquad (69)$$

Substituting the spectrum $P_c(\omega)$ (67) into relation (69), we obtain for the phase $\Delta\Phi(\omega)$

$$\Delta\Phi(\omega) = \omega T - \Phi_1(\omega), \qquad (70)$$

where the function $\Phi_1(\omega)$ represents the solution of the equation

$$\tan\Phi_1(\omega) = G(\omega)\,\tan(\omega\Delta t/2); \qquad (71)$$

$$G(\omega) = \frac{P_1(\omega) - P_2(\omega)}{P_1(\omega) + P_2(\omega)}, \qquad T = \frac{r(V_1 + V_2)}{2V_1 V_2}, \qquad \Delta t = \frac{r(V_1 - V_2)}{V_1 V_2}. \qquad (72)$$

From the phase $\Delta\Phi(\omega)$, using expression (65), it is possible to determine the velocity $V(\omega)$. Relations (70)-(72) show that the form of the dispersion function $V(\omega)$ depends on the form of the spectra $P_1(\omega)$ and $P_2(\omega)$ as well as on the relationship between the velocities V_1 and V_2. It may be inferred from (70)-(72), in particular, that for equal-power spectra $P_1(\omega)$ and $P_2(\omega)$ $[P_1(\omega) \equiv P_2(\omega)]$ the dispersion relation is absent $[V(\omega) = 2V_1V_2/(V_1 + V_2) = $ const$]$ and it is impossible to distinguish this case from the single-velocity case solely on the basis of the form of $V(\omega)$. The presence of two velocities along the line of sight in this case is felt only in the modulus of the spectrum $|P_c(\omega)|$, which has an oscillating structure with respect to the frequency ω with period $2\pi/\Delta t$ (72).

Also, since the function $|G(\omega)| \leq 1$, it therefore follows from (70), (72), and (65) that the harmonic velocity is situated between the limits V_1 and V_2:

$$V_1 \leqslant V(\omega) \leqslant V_2. \qquad (73)$$

In the low-frequency range

$$\omega \ll 2/\Delta t = 2V_1 V_2/(r\Delta V), \qquad (74)$$

it can be approximately assumed that $\Phi_1(\omega) = G(\omega)\omega\Delta t/2$, whereupon from (70) and (65) we find for $V(\omega)$

$$V(\omega) = \frac{2V_1 V_2}{(V_1 + V_2) - G(\omega)\,\Delta V}, \qquad (75)$$

i.e., the function $G(\omega)$ (72) completely determines the dispersion relation.

Inasmuch as the form of the spatial spectrum of the inhomogeneities of the interplanetary plasma is a debated issue at the moment [46-49], we examine two possibilities: Gaussian and power-law spectra $F_{N_{1,2}}^2$ (q). We first assume that the spectra $P_{1,2}(\omega)$ have a Gaussian form:

$$P_{1,2}(\omega) = C_{1,2} \exp\left(-\frac{\omega^2 a_{1,2}^2}{2V_{1,2}^2}\right), \tag{76}$$

where $a_{1,2}$ denotes the characteristic space scales of the inhomogeneities and the coefficients $C_{1,2} = P_{1,2}(0)$ characterize the powers of the spectra $P_{1,2}(\omega)$. If $V_1 > V_2$ and $V_1/a_1 > V_2/a_2$, then, as we infer from (75), the dispersion function $V(\omega)$ is a monotonically growing function for $\omega < \omega_1 = \sqrt{2}V_1V_2(a_2^2V_1^2 - a_1^2V_2^2)^{-1/2}$ and asymptotically approaches a constant level $V(\omega) = V_1$ for $\omega > \omega_1$. The harmonic velocity in this case varies from $V_0 = 2V_1V_2/[V_1 + V_2 - G(0)\Delta V]$ to V_1 for $\omega > \omega_1$, and the interval of variation of $V(\omega)$ for $C_1 > C_2$ turns out to be less than for $C_1 < C_2$. In the case $V_1/a_1 < V_2/a_2$ $(V_1 > V_2)$, the dispersion function $V(\omega)$ is monotonically decaying for $\omega < \omega_2 = i\omega_1$ and asymptotically approaches the level $V(\omega) = V_2$ for $\omega > \omega_2$. The interval of variation of $V(\omega)$ in the dispersion function $\Delta V(\omega)$ is from $V(0)$ to V_2, where $V(0) = 2V_1V_2/[V_1 + V_2 - G(0)\Delta V]$ and $\Delta V(\omega)$ is greater for $C_1 > C_2$ than for $C_1 < C_2$. We note that different situations yield different behaviors on the part of $V(\omega)$: For $V_1/a_1 > V_2/a_2$ we have a growing function $V(\omega)$ (positive dispersion), and for $V_1/a_1 < V_2/a_2$ a decaying function $V(\omega)$ (negative dispersion). These cases correspond to different signs of the cross-correlation skewness parameter: $S > 0$ in the first case and $S < 0$ in the second. This behavior of the dispersion function is realized if $\Delta V < 2(a_1/r)V_2$ (for $V_1/a_1 > V_2/a_2$) or $\Delta V < 2(a_2/r)V_1$ (for $V_1/a_1 < V_2/a_2$). Otherwise the arrival of the dispersion function at a constant level takes place at the frequency $\omega = 2/\Delta t$ (72).

We now discuss spectra $F_{N_{1,2}}^2$(q) of the power-law type

$$F_{1,2}(\mathbf{q}) = \text{const} \cdot q^{-\alpha_{1,2}}. \tag{77}$$

In the frequency interval $\omega > \omega_f = (2k/z_{1,2})^{1/2}V_{1,2}$, where k is the wave number and $z_{1,2}$ denotes the distances from the observer to the effectively scattering layers, the temporal spectra $P_{1,2}(\omega)$ (69) also have power-law form:

$$P_{1,2}(\omega) = C_{1,2}\omega^{-(\alpha_{1,2}-1)}. \tag{78}$$

We therefore investigate the dispersion function $V(\omega)$ for spectra $P_{1,2}(\omega)$ of the form (78).

Let us assume initially that $\alpha_1 = \alpha_2$. Then the function $G(\omega)$ (73) does not depend on the frequency ω $[G(\omega) = (C_1 - C_2)/(C_1 + C_2) = G_0]$, and we can use relations (66), (71)-(73) to calculate the harmonic velocity $V(\omega)$ for power-law spectra $P_{1,2}(\omega)$ (78). This dependence is shown in Fig. 3. The upper curve corresponds to the case $P_1(\omega) > P_2(\omega)$ $(V_1 > V_2, G_0 > 0)$. In the interval $\omega < 2V_1V_2/(r\Delta V)$ the dispersion relation is $V(\omega) = 2V_1V_2/(V_1 + V_2 - G_0\Delta V)$. In the interval $2/\Delta t < \omega < \pi/\Delta t$ the velocity $V(\omega)$ increases abruptly, attaining the value $V(\omega) = V_1$ for $\omega = \pi/\Delta t$; with a further increase in the frequency ω the velocity $V(\omega)$ undergoes small

Fig. 3. Dispersion function $V(\omega)$ in the model of a medium with two velocities V_1 and V_2.

$$V_1' = 2V_1V_2/(V_1 + V_2 - G_0\Delta V),$$
$$V_2' = 2V_1V_2/(V_1 + V_2 + |G_0|\Delta V),$$
$$V_0' = 2V_1V_2/(V_1 + V_2)$$

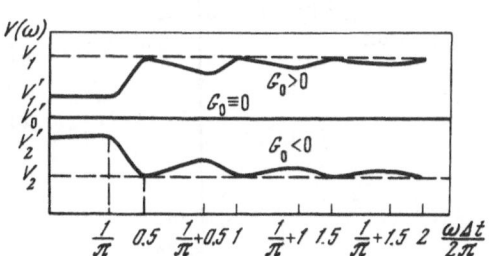

damped oscillations near the level $V(\omega) = V_1$, remaining less than V_1. The lower curve in Fig. 3 refers to the case in which $C_1 < C_2$ ($G_0 < 0$) and is the mirror image of the upper curve $V(\omega)$ about the line $V(\omega) = 2V_1V_2/(V_1+V_2)$. For $G_0 = 0$ we have $V(\omega) = 2V_1V_2/(V_1+V_2) = $ const. For $G_0 \neq 0$ the dispersion function has an asymptotic value equal to V_1 ($C_1 > C_2$) or V_2 ($C_1 < C_2$).

In experiments the frequency range of the temporal spectrum $P(\omega)$ is limited above by equipment noise ($\omega \leq \omega_{max}$), where normally ω_{max} is only a few times greater than $2V_1V_2/[r(V_1+V_2)]$ ($\omega_{max} \gtrsim 2\dot{V}_1V_2/[r(V_1+V_2)]$). Consequently, if the velocities V_1 and V_2 differ only slightly, so that $\Delta V \ll (V_1 + V_2)/2$, then the entire obserable frequency interval is $\omega < 2/\Delta t$ (see Fig. 3). In the two-velocity model with power-law spectra ($\alpha_1 = \alpha_2$) we expect a dispersion relation $V(\omega) = $ const. But if V_1 and V_2 differ considerably, $\Delta V \sim \bar{V}$, the observed function $V(\omega)$ will be constant for $\omega < 2/\Delta t$ and increase to the level $V(\omega) = V_1$ for $\omega > 2/\Delta t$.

The case $\Delta V \sim \bar{V}$ is realized in the event of accelerated flow along the line of sight, with a velocity 1.5 to 2 times the quiescent wind velocity [32].

An investigation of the scintillation temporal spectrum [12, 50] shows that the power exponent $(\alpha - 1)$ (78) varies from one day to the next. It has also been shown [50] that these variations are associated with variation of the turbulence spectrum in transition from the quiescent wind to flow, or from one flow to another. Consequently, the two-velocity model with $\alpha_1 = \alpha_2$ is not always realized. If $\alpha_1 \neq \alpha_2$, the flat parts of the curve $V(\omega)$ with $dV/d\omega = 0$ disappear. Now in the frequency interval $\omega < \pi/\Delta t$ the function $V(\omega)$ grows for $G(\omega) > 0$ and decays for $G(\omega) < 0$. The asymptotic value of $V(\omega)$ is equal to V_1 for $G(\omega) > 0$ and to V_2 for $G(\omega) < 0$.

Thus, in the two-velocity model of the medium the velocity dispersion $V(\omega)$ is a growing function, $G(\omega) > 0$, or a decaying function, $G(\omega) < 0$, in the low-frequency range $\omega < \omega_1$ and arrives at a constant level for $\omega > \omega_1$. The functions $V(\omega)$ for Gaussian and power-law spectra $P_{1,2}(\omega)$ differ very little from one another. The spread of the velocities in the dispersion function $V(\omega)$ is of the order of the differences between V_1 and V_2 [$\Delta V(\omega) \lesssim \Delta V$].

   b. E x t e n d e d   M e d i u m . We now discuss the model of a spherically symmetric solar wind with a radially directed velocity of the inhomogenieties V_0 [42, 43, 51-53]. In this model the various velocities in the diffraction pattern are associated with the dependence of the projection of the velocity $V_\perp$ onto the pattern plane on the coordinate along the line of sight (Fig. 4). The cross-correlation spectrum $P_c(\omega)$ in the weak-scintillation regime now has the form [42, 43]

$$P_c(\omega) = \int_0^\infty \frac{dz}{V_\perp(\rho)} \exp\left[i\,\frac{\omega r}{V_\perp(\rho)}\right] P_{a_s}(\omega, \rho)\, dz, \tag{79}$$

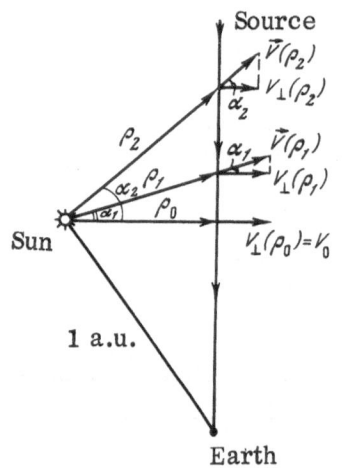

Fig. 4. Geometry of the scattering medium in the model of a spherically symmetric solar wind.

where the function $P_{a_0}(\omega, \rho)$ is proportional to the autocorrelation spectrum of the intensity fluctuations from a thin layer situated a distance z from the earth, ρ is the distance from the point of intersection of the line of sight with the layer to the sun:

$$P_{a_0}(\omega, \rho) = 8\pi A (\lambda r_e)^2 \langle \Delta n^2(\rho) \rangle \int_{-\infty}^{\infty} F_N^2 \left(q_y, q_x = \frac{\omega}{V_\perp(\rho)} \right) \sin^2 \left[\left(q_y^2 + \frac{\omega^2}{V_\perp^2} \right) \frac{z}{2k} \right] dq_y, \quad (80)$$

A is a normalization constant, $\langle \Delta n^2(\rho) \rangle$ is the local value of the mean-square value of the density fluctuations, $\rho^2 = 1 - 2z \cos \varepsilon + z^2$, ε is the elongation, and r_e is the classical electron radius. It is apparent from expression (79) that the spectrum $P_c(\omega)$ contains the characteristic frequency $\omega_c = V_0/r$, which is attributable to the oscillating factor $\exp(i\omega r/V_\perp)$.

We now ascertain the nature of the dispersion function $V(\omega)$ for the spectrum $P_c(\omega)$ (79). The velocity projection $V_\perp$ is related to the distance ρ from the sun by the expression

$$V_\perp(\rho) = V_0 \rho_0/\rho = V_0 \sin \varepsilon/\rho. \quad (81)$$

We assume that the level of the density fluctuations $\langle \Delta n^2(\rho) \rangle$ depends on the distance ρ from the sun according to the power law

$$\langle \Delta n^2(\rho) \rangle \propto \rho^{-p}. \quad (82)$$

(Observations of the scintillations have shown that the power exponent p is close to p = 4.) Then in the low-frequency range $\omega \to 0$ the harmonic velocity tends to a certain line-of-sight average $\langle V_\perp \rangle$:

$$\langle V_\perp \rangle = \frac{\int_0^\infty V_\perp^{-1}(\rho) P_{a_0}(\omega, \rho) \rho^{-p} dz}{\int_0^\infty V_\perp^{-2}(\rho) P_{a_0}(\omega, \rho) \rho^{-p} dz}. \quad (83)$$

Assuming that the form of the spectrum $F_N^2(q)$ does not depend on ρ and that the dependence of $F_N^2(q, \rho)$ on ρ is accounted for by the factor $\langle \Delta n^2(\rho) \rangle$, i.e., that the medium is mildly inhomogeneous, from (83) with the use of (81) we obtain for $\langle V_\perp \rangle$

$$\langle V_\perp \rangle = V_0 \left(\int_{-\cot \varepsilon}^{\infty} \frac{d\xi}{(1 + \xi^2)^{\frac{p-1}{2}}} \right) \left(\int_{-\cot \varepsilon}^{\infty} \frac{d\xi}{(1 + \xi^2)^{\frac{p-2}{2}}} \right)^{-1}, \quad (84)$$

whereupon for $\cot \varepsilon \gg 1$ we have

$$\langle V_\perp \rangle = \frac{2}{p-3} \left[\frac{\Gamma((p-2)/2)}{\Gamma((p-3)/2)} \right]^2 V_0 \quad (85)$$

and for several different values of p:

p	4	5	6	
$\langle V_\perp \rangle/V_0$	$2/\pi$	$\pi/4$	$8/3\pi$	(*)

It is evident from these data that the difference between $V_\perp$ and V_0 decreases with increasing value of p, which is equivalent to a decrease in the effective velocity spread.

It follows from expression (80) that the distance ρ from the sun enters into the spectrum $P_{a_0}(\omega, \rho)$ through the combination $\omega \rho/(V_0 \sin \varepsilon)$ and so the rate of decay of the function $P_{a_0}(\omega, \rho)$

with distance from $\rho = \sin \varepsilon$ increase with the frequency ω. Moreover, as the structure of the oscillating factor $\exp[i\omega r \rho/(V_0 \sin \varepsilon)]$ implies, the frequency of the z-oscillations increases with distance from $\rho = \sin \varepsilon$. As a consequence of these two effects, with an increase in the frequency ω the effective domain of the medium for the integration with respect to z in (79) becomes increasingly narrower and closer to z = cos ε. The harmonic velocity $V(\omega)$ therefore increases with the frequency ω. In the case of a Gaussian spectrum $F_N^2(\mathbf{q})$ (76) with characteristic inhomogeneity scale a greater than the baseline $r (a > r)$ the growth of the function $V(\omega)$ in the low-frequency range is induced mainly by the factor $P_{a_0}(\omega, \rho)$. Also, for $\omega > \omega_c = V_0/a$ the function $V(\omega)$ arrives at a constant level $V(\omega) = V_0$. If the inhomogeneity scale a is less than the baseline $r (a < r)$, then the arrival of $V(\omega)$ at the constant level $V(\omega) = V_0$ occurs at $\omega = \omega_c = V_0/r$. For a power-law spectrum of the inhomogeneities $F_N^2(\mathbf{q})$ (77) in the interval $\omega \gg \omega_c$ the spectrum $P_{a_0}(\omega, \rho)$ has the power-law form

$$P_{a_0}(\omega, \rho) \propto [\omega/V_\perp(\rho)]^{1-\alpha}, \tag{86}$$

since the usual baseline is of the same order of magnitude as the length of the first Fresnel zone. Now the spectrum $P_c(\omega)$ acquires the form

$$P_c(\omega) = \frac{\text{const}}{V_0} \left(\frac{\omega}{V_0}\right)^{1-\alpha} \int_0^\infty \exp\left[i \frac{\omega}{\omega_c} \frac{(1 - 2z \cos \varepsilon + z^2)^{1/2}}{\sin \varepsilon}\right] (1 - 2z \cos \varepsilon + z^2)^{-(\alpha-2)/2} dz. \tag{87}$$

To estimate the asymptotic value of the integral (87) ($\omega/\omega_c \gg 1$) we use the method of steepest descents, which gives

$$P_c^{\text{ac}}(\omega) = \frac{\text{const}}{V_0} \left(\frac{\omega}{V_0}\right)^{1-\alpha} \exp\left(i \frac{\omega r}{V_0}\right), \tag{88}$$

whence we infer that for $\omega \gg \omega_c$ the dispersion function $V(\omega) = V_0 = \text{const}$. The dispersion function $V(\omega)$ for the model of a spherically symmetric solar wind is given in Fig. 5.

Thus, in the model of an extended medium $V(\omega)$ is a growing function approaching a constant level $V(\omega) = V_0$. The low-frequency value of the velocity $\langle V_\perp \rangle$ (85) is determined by the dependence of the total power of the fluctuations $\langle \Delta n^2(\rho) \rangle$ on the distance ρ from the sun. Arrival at the constant level $V(\omega) = V_0$ takes place at the frequency $\omega_c = V_0/r$ in the case of a power-law spectrum of inhomogeneities and at the frequency $\omega_c = \min(V_0/r, V_0/a)$ in the case of a Gaussian spectrum. The usual baselines $r \sim a \sim 100$ km, making it impossible to distinguish between the power-law and Gaussian spectra according to the form of the dispersion function $V(\omega)$; the functions $V(\omega)$ corresponding to these spectra differ very little from one another. The skewness factor S is positive (S > 0) in the model of an extended medium. The velocity spread in the dispersion function $\Delta V(\omega) = V_0 - \langle V_\perp \rangle$ depends on the model of $\langle \Delta n^2(\rho) \rangle$. This spread turns out to be greater than the velocity spread determined in cross-correlation analysis. Thus, for the model of the medium with $p = 4 [\langle \Delta n^2(\rho) \rangle \propto p^{-4}]$ the velocity standard deviation (SD) $\sigma = 0.2 \langle V \rangle$. On the other hand, the velocity spread in the function $V(\omega)$, as we infer from the data [see (*) p. 107] is $\Delta V(\omega) = 0.36 V_0 = \frac{3}{8}\pi \cdot 0.36 \langle V \rangle = 0.42 \langle V \rangle$, which is twice the value

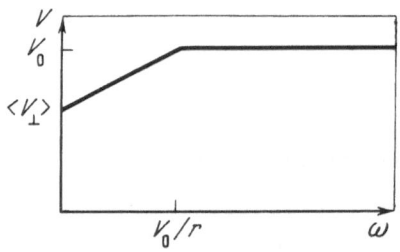

Fig. 5. Dispersion function $V(\omega)$ for the model of a spherically symmetric solar wind.

of the velocity SD. This leads to the conclusion that the dispersion function is more sensitive than the form of the cross-correlation function to the distribution of the velocities.

c. Arbitrary Velocity Distribution. Let us now assume that the velocities of the inhomogeneities are distributed according to a certain arbitrary law $\varphi(\mathbf{V})$. We consider the space-time spectrum of the intensity fluctuations:

$$P(\mathbf{q}, \omega) = \frac{1}{(2\pi)^3} \int \rho(\mathbf{r}, \tau) \exp(-i\mathbf{q}\mathbf{r} + i\omega\tau) \, d\mathbf{r} \, d\tau. \tag{89}$$

In the case of a wind scintillation pattern relation (7) is applicable to the function $\rho(\mathbf{r}, \tau)$, so that from (89) we find for $P(\mathbf{q}, \omega)$ [54]

$$P(\mathbf{q}, \omega) = 2\pi M(\mathbf{q}) \int \chi(\mathbf{q}\tau) \exp(i\omega\tau) \, d\tau, \tag{90}$$

where $\chi(\mathbf{q}\tau)$ is the characteristic function of the velocity distribution $\varphi(\mathbf{V})$. In the case of uniform motion $\varphi(\mathbf{V}) = \delta(\mathbf{V} - \mathbf{V}_0)$, $\chi(\mathbf{q}\tau) = \exp(i\mathbf{q}\mathbf{V}_0\tau)/(2\pi)^2$, and expression (90) acquires the form

$$P(\mathbf{q}, \omega) = M(\mathbf{q})\delta(\omega - \mathbf{q}\mathbf{V}_0). \tag{91}$$

From (90) we obtain an expression for the temporal cross-spectrum:

$$P_c(\omega) = \int P(\mathbf{q}, \omega) \exp(i\mathbf{q}\mathbf{r}) \, d\mathbf{q} = \frac{1}{2\pi} \int M(\mathbf{q}) \varphi(\mathbf{V}) \exp(i\mathbf{q}\mathbf{r}) \delta(\omega - \mathbf{q}\mathbf{V}) \, d\mathbf{q} \, d\mathbf{V}. \tag{92}$$

In the case of one-dimensional motion, where $\varphi(\mathbf{V}) = \varphi(V_x)\delta(V_y)$, the spectrum $P_c(\omega)$ takes the form

$$P_c(\omega) = \int \frac{1}{V} \exp\left(i\frac{\omega r}{V}\right) P_a(\omega, V) \varphi(V) \, dV, \tag{93}$$

where $P_a(\omega, V)$ is the autocorrelation spectrum of uniform motion with velocity V:

$$P_a(\omega, V) = \int M(\mathbf{q}) \delta(q_x - \omega/V) \, d\mathbf{q}. \tag{94}$$

Let us consider the dispersion function $V(\omega)$ in the low-frequency range ($\omega \to 0$). As $\omega \to 0$ the spectrum $P_a(\omega, V)$ (94) tends to a constant value

$$P_a(0, V) = \int M(q_x = 0, q_y) \, dq_y. \tag{95}$$

We therefore infer from (93) that

$$V(0) = \frac{\langle V^{-1} \rangle}{\langle V^{-2} \rangle} = \int \frac{1}{V} \varphi(V) \, dV \cdot \left[\int \frac{1}{V^2} \varphi(V) \, dV\right]^{-1} \tag{96}$$

and at low frequencies ($\omega \to 0$) the harmonic velocity is less than the average velocity of the inhomogeneities, because

$$\langle V^{-1} \rangle < \langle V^{-2} \rangle \langle V \rangle. \tag{97}$$

We now investigate the dispersion function $V(\omega)$ at frequencies $\omega \neq 0$. The integrand in (93) represents the product of a slowly varying function $P_a(\omega, V)\varphi(V)$ and an oscillating member exp $(i\omega r/V)$. For a fixed frequency ω the function $P_a(\omega, V)$ increases with the value of V, the rate of growth increasing with the frequency ω. In addition, at ω = const the oscillation frequency of the factor exp $(i\omega r/V)$ increases with decreasing value of V. Consequently, for sufficiently narrow distribution functions $\varphi(V)$ (velocity SD much less than the average velocity of the inhomogeneities: $\sigma \ll \langle V \rangle$) the dispersion function is a growing function. The actual distribution $\varphi(V)$ must be bounded with respect to the velocity V [$\varphi(V)$ = 0 for V > V_{max}], and so the dispersion function $V(\omega)$ arrives at a plateau $V(\omega)$ = V_{max} for ω > V_{max}/r. This behavior of $V(\omega)$ is reminiscent of the dispersion function in the extended-medium model. However, these two models differ in the variation of $V(\omega)$ with the elongation ε; in the extended-medium model the velocity spread $\Delta V(\omega)$ increases with the value of ε, while in the model with random velocities, clearly, the opposite trend will occur due to the depreciation of the random velocity component as ε increases [38]. For large elongations ε we are interested in the case of small velocity SD ($\sigma \ll \langle V \rangle$), where the dispersion function is a growing function.

If the velocity SD σ is large ($\sigma \gtrsim \langle V \rangle$; this case can be realized for small values of ε [38]), then the dispersion function can become a decreasing function. For the inception of negative dispersion $dV/d\omega < 0$ it is necessary that the crowding of the oscillations in the integral (93) with increasing value of ω be more influential at larger velocities V > $\langle V \rangle$. This situation is possible for $\sigma \sim \langle V \rangle$ in the case of distributions $\varphi(V)$ that either have negative skewness ($\gamma_1 < 0$) or are flatter than normal ($\gamma_2 < 0$). For such distributions, as was shown in Section 1, the cross-correlation skewness (37) is negative (S < 0). Accordingly, it may be inferred that negative dispersion ($dV/d\omega < 0$) is possible only for distributions $\varphi(V)$ with the moments related in such a way as to ensure negative cross-correlation skewness.

As an example we discuss a Gaussian scintillation spectrum $M(q) \propto$ exp $(-q^2 a^2/2)$ and a normal velocity distribution $\varphi(V)$ (9). In this case relation (93) takes the form

$$P_c(\omega) = \int_{-\infty}^{\infty} \frac{1}{V} \exp\left[-\frac{\omega^2 a^2}{2V^2} - \frac{(V - \langle V \rangle)^2}{2\sigma^2} + i\frac{\omega r \cos\theta}{V}\right] dV, \tag{98}$$

where θ is the angle between the velocity $\langle \mathbf{V} \rangle$ and the baseline $\mathbf{r}$. The value of $V(\omega)$ for $\omega \to 0$, when exp $(-\omega^2 a^2/2V^2)$ = const, has been obtained above [see (96)]. We wish to determine the behavior of $V(\omega)$ for $\omega \gtrsim \langle V \rangle/a$. In (98) we make the change of variables $1/V = \xi$, whereupon we obtain for $P_c(\omega)$

$$P_c(\omega) = \int_{-\infty}^{\infty} \frac{d\xi}{\xi} \exp\left[-\frac{\omega^2 a^2 \xi^2}{2} - \frac{(\langle V \rangle \xi - 1)^2}{2\sigma^2 \xi^2} + i\omega r \xi \cos\theta\right]. \tag{99}$$

In the analyzed frequency range $\omega \gtrsim \langle V \rangle/a$ expression (99) comprises the Fourier transform of the function

$$\Phi(\xi) = \frac{1}{\xi} \exp\left[-\frac{\omega^2 a^2 \xi^2}{2} - \frac{(\langle V \rangle \xi - 1)^2}{2\sigma^2 \xi^2}\right], \tag{100}$$

which vanishes as $\xi \to 0$ and as $\xi \to \infty$ and attains a maximum for a certain $\xi = \xi_0$. The condition for a maximum of $\Phi(\xi)$ has the form

$$\omega^2 a^2 \sigma^2 \xi_0^4 + \langle V \rangle \xi_0 - 1 = 0. \tag{101}$$

For $\sigma = 0$ expression (101) yields $\xi_0 = 1/\langle V \rangle$ (for uniform motion). In the frequency interval $\omega \gtrsim \langle V \rangle/a$ we obtain from (101) for $\sigma \ll \langle V \rangle$

$$\xi_0 = \frac{1}{\langle V \rangle}\left(1 - \frac{\omega^2 \sigma^2}{\omega_0^2 \langle V \rangle^2}\right), \qquad \omega_0 = \frac{\langle V \rangle}{a}. \tag{102}$$

The spectrum $P_c(\omega)$ for $\omega \gtrsim \langle V \rangle/a$ according to the properties of the Fourier transform of shifted functions can be represented by the product of an amplitude factor $W(\omega)$ and a phase factor $\exp(i\omega r \cos\theta\xi_0)$. Now from relation (66) we obtain

$$V(\omega) = \langle V \rangle \left(1 + \frac{\omega^2\sigma^2}{\omega_0^2\langle V\rangle^2}\right); \qquad \sigma \ll \langle V \rangle. \tag{103}$$

Analogous reasoning for $\sigma \gg \langle V \rangle$, in which case the third term in (101) can be neglected, results in the function

$$V(\omega) = (\omega a\sigma)^{1/2}, \qquad \sigma \gg \langle V \rangle. \tag{104}$$

The function $V(\omega)$ for $\sigma \gg \langle V \rangle$ (104) and for $\omega \ll \langle V \rangle$ (103) is given in Fig. 6.

In the models of the medium a-c discussed above it has been assumed that the inhomogeneities do not vary with time and are transported by the solar wind, i.e., that the homogeneities are locally frozen-in and the IPS pattern has a wind nature. The dispersion function $V(\omega) \neq$ const in this case is not a characteristic of the true dispersion character of the motion of the inhomogeneities, rather it characterizes the structure of the velocities in the scattering medium. The velocity spread in the dispersion function is of the order of the velocity spread in the medium. It is difficult to distinguish between the different models of the medium on the basis of the form of the dispersion function, because the disparities between $V(\omega)$ in the different models are slight. Consequently, in order to determine the nature of the observed velocity spread it is necessary to use the function $V(\omega)$ in conjunction with an analysis of the variations of $V(\omega)$ with the elongation and heliolatitude.

d.   Time Variation of Inhomogeneities. Let us now suppose that the local frozen-in character of the inhomogeneities is violated, i.e., that the inhomogenities vary with time, and let us see how these variations affect the observed parameters of the IPS pattern. To simplify matters we confine the discussion to the model of a thin phase screen. When the frozen-in character is lost, it is necessary to consider the space-time fluctuations of the density of scattering particles $\Delta N(\mathbf{r}, t)$, which corresponds to the space-time turbulence spectrum $F_N^2(\mathbf{q}, \omega)$. If the time rate of change of the inhomogeneities in this case is much less than the velocity of light, then in a frame moving with the solar wind the space-time scintillation spectrum is

$$M(\omega, \mathbf{q}) = 4\sin^2\left(q^2\frac{z}{2k}\right) \cdot F_N^2(\mathbf{q}, q_z = 0, \omega). \tag{105}$$

Transforming to a fixed frame, we obtain from (105) for the auto- and cross-correlation spectra of the scintillations:

$$P_a(\omega) = \int M(\mathbf{q}, q_z = 0, \omega - \mathbf{q}\mathbf{V}_0)\, d\mathbf{q}, \tag{106}$$

$$P_c(\omega) = \int M(\mathbf{q}, q_z = 0, \omega - \mathbf{q}\mathbf{V}_0)\exp(i\mathbf{q}\mathbf{r})\, d\mathbf{q}. \tag{107}$$

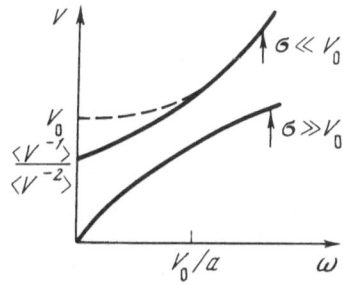

Fig. 6. Dispersion function $V(\omega)$ for a normal velocity distribution.

We recall that in the case of frozen inhomogeneities the temporal autocorrelation spectrum $P_a(\omega)$ (94) has the form $P_a(\omega) = \int M(\mathbf{q})\delta(\omega - q V_0)d\mathbf{q}$. A comparison of (106) and (94) shows that allowance for time variation of the inhomogeneities broadens the spectrum $P_a(\omega)$ by comparison with the case of frozen motion.

Within the scope of the hypothesis of turbulence of the interplanetary plasma [55-59] it is assumed that the density fluctuations in the solar wind are associated with low-frequency plasma turbulence. If the turbulence is mild, the broadening of the resonances [60] can be neglected, whereupon the frequency ω is related to the spatial frequency $\mathbf{q}$ by the dispersion relation $\omega = \omega(\mathbf{q})$. The space-time spectrum of the density fluctuations now has the form

$$F_N^2(\mathbf{q}, \omega) = \frac{1}{2} F_N^2(\mathbf{q})\{\delta[\omega - \omega(\mathbf{q})] + \delta[\omega + \omega(\mathbf{q})]\}. \tag{108}$$

Let us consider the temporal spectrum $P_c(\omega)$ and the dispersion function $V(\omega)$ for the plasma wave modes potentially responsible for interplanetary scintillations.

1. $\omega(\mathbf{q}) = (\mathbf{q}\mathbf{V}_0/V_0)\tilde{V}$, where V_0 is the wind velocity and $\tilde{V}$ is the phase velocity of the waves. This type of dispersion relation is realized for waves propagating in a selective direction coinciding with the solar wind velocity $\mathbf{V}_0$. The role of this direction can be taken by the direction of the magnetic field in the case of Alfvén waves or magnetized ion-acoustic waves. The substitution of (108) into (107) and (106) shows that the given situation is reducible to the two-velocity case (see Section 2.2a). The dispersion function is a growing function and is described by expression (75) with the substitution $V_1 = V_0 + \tilde{V}$; $V_2 = V_0 - \tilde{V}$; $G(\omega) = (P_1 - P_2)/(P_1 + P_2)$, $P_{1,2}(\omega) = \int M(q_x = \omega(V_0 \pm \tilde{V}), q_y)dq_y$ for $\tilde{V} \ll V_0$. If the phase velocity $\tilde{V}$ becomes of the same order as V_0, the function $V(\omega)$ acquires oscillations, and the cross-correlation function $\rho(\mathbf{r}, \tau)$ acquires two maxima (instead of one): $\tau_{01} = r/(V_0 + \tilde{V})$ and $\tau_{02} = r/(V_0 - \tilde{V})$, where these maxima are characteristically of the same magnitude. The model of a turbulent medium with waves traveling in two directions can be realized for a local turbulence source. The emergence of two maxima and an oscillating behavior on the part of $V(\omega)$ is possible only in regions close to the sun, because the plasma wave velocity increases with nearness to the sun.

2. $\omega(\mathbf{q}) = q_y\tilde{V}$. This relation holds in the case of waves traveling predominantly transversely with respect to $\mathbf{V}_0$, as for example in the case of magnetoacoustic waves, which propagate mainly across the magnetic field $\mathbf{B}(\mathbf{V}_0 \parallel \mathbf{B}$ for $\rho < 1$ a.u.) with a phase velocity $\tilde{V} = V_A = B/(4\pi MN)^{1/2}$. In this case, as implied by (107) and (108), there is no dispersion dependence for $\mathbf{V}_0 \parallel \mathbf{B}$ [$V(\omega) = V_0 + \tilde{V}^2/V_0 = \text{const}$], and the value of $V(\omega)$ coincides with the value of the apparent velocity V_a of the diffraction pattern. The cross-correlation skewness factor S is equal to zero for any orientation of the baseline $\mathbf{r}$ with respect to the velocity $\mathbf{V}_0$. It is important to note that with an increase in the velocity $\tilde{V}$ the correlation $\rho(\mathbf{r}, \tau_0)$ at the maximum will decrease, and for $r\tilde{V} > aV_0$ (a is the inhomogeneity scale) the correlation of the scintillations between stations practically vanishes.

3. $\omega(\mathbf{q}) = q\tilde{V}$. This relation holds in the case of isotropic turbulence, as for example in connection with nonmagnetized ion-acoustic waves ($\tilde{V} = V_S = (T_e/M)^{1/2}$. For isotropic waves we obtain from (107) and (108)

$$P_c(\omega) = \frac{1}{V_0}\exp\left(i\frac{\omega r}{V_0}\right)\left[\int_{\omega|(V_0 - \tilde{V})}^{\infty} M(q)\exp\left(iqr\frac{\tilde{V}}{V_0}\right)dq + \int_{\omega|(V_0 + \tilde{V})}^{\infty} M(q)\exp\left(-iqr\frac{\tilde{V}}{V_0}\right)dq\right], \tag{109}$$

whence it follows that the dispersion function is a growing function. Thus, it can be inferred from (109) that as $\omega \to 0$ the function $V(\omega) \to V_0$ and as $\omega \to \infty$ the function $V(\omega) \to V_0 + \tilde{V}$. If $\tilde{V} \ll V_0$, the velocity spread in the dispersion relation is inconsequential; if $\tilde{V} \sim V_0$, the spread $\Delta V(\omega)$ becomes large, but in this case the correlation of the scintillations between stations falls off abruptly.

4. $\omega(\mathbf{q}) = \omega_0 = \text{const.}$ This type of dispersion relation is realized in the case of turbulence at the natural frequencies of the plasma, for example in connection with ion-cyclotron modes $\omega_0 = \omega_{B_i} = eB/(Mc)$. The spectrum $P_c(\omega)$ (107) in this case assumes the form

$$P_c(\omega) = \frac{1}{V_0}\left[P_a^-(\omega)\exp\left(i\,\frac{\omega-\omega_0}{V_0}r\right) + P_a^+(\omega)\exp\left(i\,\frac{\omega+\omega_0}{V_0}r\right)\right], \qquad (110)$$

where $P_a^\pm(\omega) = \int M(q_x = (\omega \pm \omega_0)/V_0,\, q_y)dq_y$. If the spectrum $P_a(\omega)$ has a Gaussian form, $P_a(\omega) \propto \exp(-\omega^2 a^2/2V_0^2)$ (a is the inhomogeneity scale), then the phase $\Delta\Phi(\omega)$ of the cross-spectrum $P_c(\omega)$ (110) is

$$\Delta\Phi(\omega) = \omega r/V_0 + \Phi_1(\omega), \qquad (111)$$

where

$$\Phi_1(\omega) = \arctan\left[\tanh\frac{\omega\omega_0 a^2}{V_0^2}\cdot\tan\frac{\omega_0 r}{V_0}\right]. \qquad (112)$$

It is evident from (111) and (112) that for $\omega_0 r/V_0 < \pi/2$ the dispersion function $V(\omega)$ is a growing function: $V(\omega) = V_0(1 + \omega_0^2 a^2/V_0^2)^{-1}$ as $\omega \to 0$ and $V(\omega) = V_0\omega/(\omega + \omega_0)$ as $\omega \to \infty$. The velocity spread in the function $V(\omega)$ increases with the oscillation frequency ω_0. For $\pi/2 < \omega_0 r/V_0 < \pi$ the dispersion function becomes a decreasing function: $V(0) = V_0(1 - \omega_0^2 a^2/V_0^2)^{-1}$ and $V(\omega \to \infty) = V_0\omega/(\omega - \omega_0)$. In the interplanetary medium the frequency ω_0 varies with the elongation: $\omega_0 \propto \rho^{-2}$. Consequently, on moving away from the sun alternating regimes with $dV/d\omega > 0$ and $dV/d\omega < 0$ are encountered. At a sufficient distance from the sun, such that ω_0 becomes less than $\pi V_0/2r$, the regime with $dV/d\omega > 0$ stabilizes. This behavior on the part of $(dV/d\omega)\rho$ provides a tool, in principle, for the detection of the given oscillation mode. It is important to note that close to the sun, at distances ρ for which $\omega_0 > V_0/a$ [in the case of ion-cyclotron oscillations this region is one of subsonic flow, since $a \sim (\omega_0/V_{Ti})^{-1} \sim \rho_i$ is the Larmor ion radius], the temporal autocorrelation spectrum $P_a(\omega)$ (106) acquires a maximum at the frequency $\omega = \omega_0$ rather than at $\omega = 0$ for other types of turbulence. The occurrence of such a maximum at small elongations would be evidence in favor of ion-cyclotron modes.

Thus, time variation of the inhomogeneities, like the structure of the velocities in the scattering medium, can induce a dependence $V(\omega)$. In models of the medium with plasma turbulence entrained by the solar wind the scatter of the velocities is of the order of the characteristic velocities in the plasma [velocity of sound $V_S = (T_e/M)^{1/2}$, Alfvén velocity $V_A = B/(4\pi NM)^{1/2}$]. For large elongations $\rho \gtrsim 0.5$ a.u., where IPS observations are generally made, the velocity spread $\Delta V(\omega)$ associated with entrained turbulence is not greater than 100 km/sec, which is less than the spread $\Delta V(\omega)$ associated with the velocity structure. As the sun is approached the characteristic velocities and frequencies associated with plasma turbulence increase. In this case the above-considered singular attributes of the functions $V(\omega)$ and spectra $P_a(\omega)$, typical of the various types of turbulence, must be manifested. The detection and analysis of such attributes make it possible in principle to draw definite conclusions regarding the nature, not only of the velocity spread, but also of the inhomogeneities themselves. These considerations place vital importance on the investigation of scintillations at several stations in the regions close to the sun.

2.3. Correspondence between the Dispersion and

Cross-correlation Methods of Scintillation Analysis

We now inquire how cross-correlation effects of the scintillation process (broadening and skewing) are related to the dispersion function $V(\omega) \neq \text{const.}$

We first consider the special case in which the condition for applicability of the BPS method is satisfied (see Section 2.1), i.e., the correlation cross section $\rho(x, \tau)$ has the form

$$\rho\,(x,\tau) = \rho\,[x^2 - 2x\,\langle V \rangle \tau + (\langle V \rangle^2 + \sigma^2)\tau^2], \tag{113}$$

where σ is the standard deviation (SD) of the velocities. The correlation cross section (113) is symmetric about the maximum of $\rho(x, \tau)$, which is attained for $\tau = \tau_0 = x/V_a$, where $V_a = \langle V \rangle + \sigma^2/\langle V \rangle$ is the apparent velocity of the IPS pattern. It can be verified that the dispersion function $V(\omega)$ for the cross-spectrum $P_c(\omega)$ corresponding to the cross section (113) is absent: $V(\omega) = \langle V \rangle + \sigma^2/\langle V \rangle = \text{const}$. A spectral analog of the BPS method has been proposed [61] on the basis of the latter fact and entails the following. From the known modulus $|P_c(\omega)|$ and phase $\Delta\Phi(\omega)$ of the spectrum it is possible to determine the apparent velocity of the diffraction pattern and the average velocity $\langle V \rangle$. To do so it is required to construct the surface $|P_c(\omega, \Delta\Phi)|$ and to delineate on it the contours of constant spectral power $|P_c(\omega, \Delta\Phi)| = \text{const}$ (Fig. 7). Inasmuch as $\Delta\Phi = qx$ for uniform motion, the condition $\Delta\Phi = \text{const}$ is equivalent to $q = \text{const}$. Here it is possible to draw two lines through the cross sections of constant spectral power; one of them corresponds to the maximum amplitude for a fixed wave number q, and the second to the maximum amplitude $|P_c(\omega)|$ for a fixed frequency ω (Fig. 7). Two characteristic velocities can be determined from these lines: $V_1 = x \tan \alpha_1$, $V_2 = x \tan \alpha_2$, $V_1 > V_2$ (Fig. 7). The application of this method and the BPS method to the same laboratory drift process [61] has shown that the velocity V_1 is close to the apparent velocity V_a, and the velocity V_2 to the average velocity of the inhomogeneities $\langle V \rangle$. It also turns out that the frequency dependence of the phase $\Delta\Phi(\omega)$ of the cross-spectrum $P_c(\omega)$ is close to linear and the pattern velocity obtained from the function $\Delta\Phi(\omega)$ $[V_3 = \omega x/\Delta\Phi(\omega)]$ coincides with the apparent velocity $(V_a \simeq V_3)$. Thus, for the correlation cross section (113) we have established a correspondence between the BPS method and the spectral method.

However, the spectral method described above is bound by the same restrictions as the BPS method. It is expected, therefore, that its application to IPS will be attended by the same difficulties as the BPS method. Thus, it has been shown above that allowance for the fine structure of the solar-wind velocities leads to a nonlinear dependence of the phase $\Delta\Phi(\omega)$ on the frequency ω. As a result, even for a slight velocity dispersion the constant-spectral-power contours will deviate from elliptical. These deviations are the most pronounced at low frequencies $\omega < V_a/x$.

We now examine the general case $V(\omega) \neq \text{const}$ and determine how the existence of the dependence $V(\omega)$ is related to the parameters of the cross-correlation function $\rho(x, \tau)$:

$$\rho\,(x,\tau) = \int\limits_{-\infty}^{\infty} |P_c(\omega)| \exp\left[i\,\frac{\omega x}{V(\omega)}\right] \exp\,(i\omega\tau)\,d\omega. \tag{114}$$

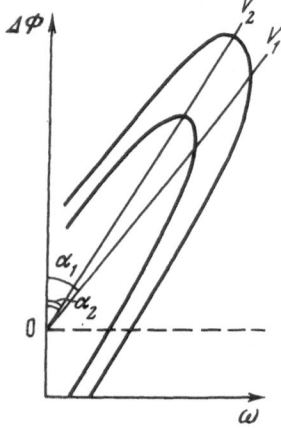

Fig. 7. Contours of constant spectral power $|P_c(\omega, \Delta\Phi)| = \text{const}$ in the plane $(\omega, \Delta\Phi)$.

We consider the parameter τ_2, i.e., the width of the function $\rho(x, \tau)$ (114), defining it as the reciprocal of the second moment of the temporal spectrum $P_c(\omega)$. Then according to (24) we have

$$\tau_2^2 = \frac{\int |P_c(\omega)| \exp[i\omega(x/V(\omega) - \tau_0)]\, d\omega}{\int |P_c(\omega)| \exp[i\omega(x/V(\omega) - \tau_0)]\, \omega^2 d\omega}. \tag{115}$$

It is evident from (115) that in the case of uniform motion, where $V(\omega) = \text{const}$ and $|P_c(\omega)| = P_a(\omega)$, the width τ_2 of the cross-correlation function coincides with the width of the autocorrelation function: $\tau_1 = \tau_2$ ($x = 0$, $\tau_0 = 0$). But if different velocities of the inhomogeneities are present along the line of sight, $V(\omega) \neq \text{const}$, then for weak dispersion $V(\omega) = V_a + \Delta V(\omega)$ [$\Delta V(\omega) \ll V_a$] we obtain from (115)

$$\tau_2^2 - \tau_1^2 = \frac{1}{2} \tau_1^2 \frac{x^2}{V_a^4}\left[\frac{\langle \omega^4 \Delta V^2(\omega)\rangle}{\langle \omega^2 \rangle} - \langle \omega^2 \Delta V^2(\omega)\rangle\right], \tag{116}$$

where $\langle f(\omega)\rangle = \int f(\omega) P_a(\omega) d\omega [\int P_a(\omega) d\omega]^{-1}$, $V_a = x/\tau_0$. A comparison of the terms in the brackets of (116) leads to the conclusion that $\tau_2 > \tau_1$ and so the presence of the dispersion function $V(\omega)$ induces broadening of the cross-correlation function $\rho(x, \tau)$ relative to the autocorrelation function $\rho(0, \tau)$.

We now inquire how the nature of the function $V(\omega)$ affects the cross-correlation skewness parameter S. Substituting $\rho(x, \tau)$ (114) into the definition of S (34) and taking into account the maximum condition $\partial \rho(x, \tau)/d\tau|_{\tau_0} = 0$, we find

$$S = \int_{-\infty}^{\infty} \sin \omega t \cdot \sin\left[\frac{\omega x}{V_a}\left(\frac{V_a}{V(\omega)} - 1\right)\right] \cdot P_a(\omega)\, d\omega, \tag{117}$$

whereupon, using the relation $V(\omega) = V(-\omega)$, we have

$$S = 2 \int_0^{\infty} \sin \omega t \cdot \sin\left[\frac{\omega x}{V_a}\left(\frac{V_a}{V(\omega)} - 1\right)\right] \cdot P_a(\omega)\, d\omega. \tag{118}$$

The main contribution to the integration of (118) is from low frequencies $\omega < V_a/a$, so that the skewness parameter S is positive for $V(\omega) < V_a$ ($dV/d\omega > 0$) and negative for $V(\omega) > V_a$ ($dV/d\omega < 0$), i.e.,

$$\text{sgn}\, S = \text{sgn}\,(dV/d\omega). \tag{119}$$

Relations (116) and (118) establish a correspondence between the distortion of the function $\rho(x, \tau)$ and the function $V(\omega)$.

By analyzing the cross-correlation skewness and broadening it is posible to detect the presence of velocity fine structure and to estimate the velocity dispersion. Investigation of the function $V(\omega)$ provides a means for analyzing the nature of the observed velocity spread. However, this approach incurs two difficulties. First, it is impossible on the basis of individual functions $V(\omega)$ to distinguish between "wind" and "vorticity" patterns (velocity fine structure or time variation of the inhomogeneities). Second, the different models of the velocity structure in the solar wind do not differ too significantly in the nature of the function $V(\omega)$, and it is difficult to choose between the different models solely on the basis of individual functions $V(\omega)$. In analyzing the observed function $V(\omega)$, therefore, it is necessary to recruit data obtained from satellite measurements of the velocity distribution and to allow for variation of $V(\omega)$ and other scintillation parameters with the elongation and heliolatitude. It is essential to note, however, that the indicated difficulties are inherent in all methods of investigation of the velocities according to scintillation observations at several stations. They are attributable to the specifics of the temporal measurements.

Three facts tend to make attractive the idea of using dispersion analysis for investigation of the velocity fine structure. First, the nature of the dispersion function is practically independent of the form of the spatial spectrum of the inhomogeneities. Consequently, for analysis of the velocity fine structure by the dispersion method it is inessential to know the form of the spectrum $F_N^2(q)$. Second, the spread of the velocities in the dispersion function $V(\omega)$ is greater than the effective spread of the velocities in the cross-correlation method, i.e., dispersion analysis is more sensitive to the velocity spread than the cross-correlation method. Third, on the basis of the function $V(\omega)$ it is possible to obtain information in pure form about the actual velocities present in the medium, in particular about the higher velocities, whereas the parameters of the cross-correlation function are associated strictly with the integral characteristics of the velocity distribution function and can be quite insensitive to the "tails" of the distribution function (V).

In analyzing the velocity spread it is necessary to consider the orientation of the baseline relative to the velocity vector, i.e., $\cos\theta$ in (65), and so for the implementation of dispersion analysis it is desirable to have at least three stations on earth.

2.4. Selection of Optimal Baselines for Dispersion Analysis

The values of the harmonic velocities in the dispersion function are always determined with a certain error $\delta V(\omega)$:

$$\delta V(\omega) = V(\omega)\delta\Phi(\omega)/\Phi(\omega), \tag{120}$$

where $\delta\Phi(\omega)$ is the error in the determination of the phase spectrum $\Phi(\omega)$ due to the finiteness of the observation period T. Consequently, the use of individual functions $V(\omega)$ for investigation of the velocity structure is admissible only with the provision that the velocity variation in the dispersion function is greater than δV (120):

$$\Delta V(\omega) > \delta V(\omega). \tag{121}$$

On the other hand, an analysis of various models of the velocity structure of the solar wind shows that the function $V(\omega)$ becomes steeper with increasing baseline r in the frequency interval where $\Delta\Phi(\omega) < 1$. This means that for a fixed frequency ω [$\Delta\Phi(\omega) < 1$] the value of $\Delta V(\omega) = V(\omega) - V(0)$ increases with r. Consequently, to guarantee condition (121) it is necessary either to increase the baseline r or to increase the observation time T.

We therefore discuss the selection of the optimal parameters r and T. An integral characteristic of the dispersion function $V(\omega)$ is the cross-correlation skewness parameter S (37), $S \sim (r/a)(\sigma^2/\langle V\rangle^2)$. We calculate the error in S due to the finiteness of the interval $T \neq \infty$. The estimator of the correlation function $\rho(\mathbf{r}, \tau)$ calculated over the interval T has the form

$$\hat{\rho}(\mathbf{r}, \tau) = \frac{1}{T}\int_0^T \delta I(0, t)\delta I(\mathbf{r}, t + \tau)\, dt \tag{122}$$

and differs from the correlation function $\rho(\mathbf{r}, \tau) = \langle\delta I(0, t)\delta I(\mathbf{r}, t + \tau)\rangle$ by a certain random quantity $\Delta\rho(\tau)$, the variance of which is equal to

$$\langle\Delta\rho^2(\tau)\rangle = \frac{1}{T^2}\iint [\rho^2(0, t - t') + \rho(\mathbf{r}, t - t' - \tau)\rho(\mathbf{r}, t' - t - \tau)]\, dt\, dt'. \tag{123}$$

From relation (123) we find $\langle \Delta\rho^2(\tau) \rangle$ for $\rho(0, \tau) = \exp(-\tau^2/2\tau_1^2)$ and $\rho(\mathbf{r}, \tau) = \exp[-(\tau - \tau_0)^2/(2\tau_1^2)]$:

$$\langle \Delta\rho^2(\tau) \rangle = \frac{\sqrt{\pi}\,\tau_1}{T}[1 + \rho(0, \sqrt{2}(\tau_0 + \tau))]. \tag{124}$$

Using the definition of the skewness parameters S (34) and relation (124) for $\langle \Delta\rho^2(\tau) \rangle$, we calculate the variance of the skewness S for $t \sim \tau_1$:

$$\langle \delta S^2 \rangle = \langle (S - \langle S \rangle)^2 \rangle = 2\sqrt{\pi}\tau_1/T. \tag{125}$$

Condition (121) is equivalent to the condition

$$S \sim \frac{r}{a} \frac{\sigma^2}{\langle V \rangle^2} > 3\,(\langle \delta S^2 \rangle)^{1/2} = 3\left(\frac{2\sqrt{\pi}\,\tau_1}{T}\right)^{1/2}, \tag{126}$$

from which, for a fixed baseline r, we find the following constraint on the length of the interval T:

$$T > T_{1m} = 18\sqrt{\pi}\tau_1\,(\langle V \rangle/\sigma)^4(a/r)^2. \tag{127}$$

Assuming for interplanetary scintillations that $\tau_1 \sim 0.5$ sec, $(\langle V \rangle/\sigma)^4 \sim 10^2$, $a \sim 150$ km, and specifying the values of the baseline r, we estimate the values of the minimum interval T_{1m} necessary for the implementation of dispersion analysis:

r, km	50	100	200	450	600	1500	2000
T_{1m}, min	240	60	15	3	1.5	0.25	0.16

It is evident from these data that the value of T_{1m} falls off rapidly with increasing baseline r. Consequently, for dispersion analysis with an interval $T \sim 10$ min fairly long baselines are required ($r > 200$ km). It is important to remember, however, that upper bounds also exist for the baseline r. One of those bounds is associated with transverse drift of inhomogeneities, which induces a loss of correlation approximately according to the law $\exp(-\frac{1}{2}r^2 \sin^2\theta/a^2)$. Therefore, baselines $r > 200$ km are suitable only for drift directed along the baseline. Another bound is imposed by rearrangement of the diffraction pattern. As a result of the latter effect a loss of IPS correlation takes place approximately according to the law $\exp[-r^2\sigma^2/(a^2\langle V \rangle^2)]$. Hence it follows that baselines with $r > a\langle V \rangle/\sigma \sim 4a \sim 600$ km are totally unstable for investigation of the inhomogeneity velocities.

Thus, the optimal baseline for investigation of the velocity fine structure have lengths $r \simeq 200$ km. The observation time for such baselines must be greater than 15 min. For a baseline $r \simeq 100$ km the observation time must be of the order of one hour.

3. Experimental Data

Multiple-station scintillation observations had their origin in the Laboratory of Radio Astronomy (LRA) of the P. N. Lebedev Physics Institute of the Academy of Sciences of the USSR where the first measurements of the velocity vector of the solar wind were performed [62]. These investigations were soon thereafter pursued vigorously in other countries, including Great Britain [4, 63], the United States [12, 18, 64], and Japan [8, 65]. The points at which the measurements of the intensity fluctuations were performed were the vertices of a triangle with sides of ~100 to 200 km [4, 8, 18, 21]. Originally in these studies the apparent velocity of the scintillation pattern and the inhomogeneity space scale were determined. It turned out that the apparent velocity agrees quite well with the solar-wind velocity, i.e., with the proton velocity measured on spaceships [66]. The correlation of the intensity at stations separated by a distance of $\simeq 200$ km with average pattern drift along the baseline attained 0.7 to 0.9 [66]. Such a high degree of correlation indicates that the lifetime of the diffraction pattern is large in comparison with its lag time ($\tau_c \gg \tau_0$) and so the role of diffusion spreading of the inhomogeneities is slight. This fact exposes a possibility for analysis of the velocity structure of the solar wind.

3.1. Cross-correlation Analysis

The problem of the statistical variance of the velocities in the solar wind was first formulated in [20, 38]. In [38] the standard deviation (SD) of the velocities of the inhomogeneities was measured on the basis of the LE method from observations of scintillations of the radio source 3C279 at two stations separated by a distance r = 16 km, at high frequencies f = 2295 MHz. It was found that in the interval of small elongations $10R_\odot < \rho < 40R_\odot$ the velocity SD increases the closer the radio source is to the sun, attaining a maximum value $\sigma_{max} \simeq 150$ km/sec in the vicinity of $R \simeq 10R_\odot$. The average velocity of the inhomogeneities in this case begins to diminish appreciably with nearness to the sun for $\rho < 30R_\odot$. This dependence of the solar-wind velocity on the distance ρ from the sun is consistent with Parker's model of an expanding solar corona.

Golley and Denisson [39] undertook the first attempt to estimate the cross-correlation skewness parameter S from three-station observations of the source 3C48 at f = 81.5 MHz with baselines of 85, 66, and 52 km. The average skewness parameter S (46) was determined in a 23-day observation period in 1966: S = 0.035 ± 0.008. In this study the skewness S was interpreted as an indicator of velocity spread of the inhomogeneities in the medium.

Lotova and Vereshchagina [67] investigated the cross-correlation skewness in greater detail, determining S from three-station observations in 1967-69 at the Radio Astronomy Station (RAS) of the Physics Institute, using the sources 3C38 and 3C144 at a frequency of 86 MHz. The behavior of the skewness was compared with the behavior of the other parameters characterizing IPS — the scintillation index m, the apparent velocity V_a, and the apparent space scale a_2 — and it was shown that the variations of S correlate with the variations of m, V_a, and a_a. It was also shown that the velocity fine structure is a constant property of the IPS pattern, and the variations of the parameter S are determined by the solar-wind dynamics.

Using the values found for S in [39], we determine the velocity SD σ in the solar wind. Substituting the parameters r $\simeq$ 70 km, $a \simeq 160/\sqrt{2}$ km, and $\delta \simeq 1$ into expression (51), we obtain the relative velocity SD $\sigma/\langle V \rangle$:

$$\frac{\sigma}{\langle V \rangle} \simeq \left(S \frac{a}{r} \right)^{\frac{1}{2}} \simeq 0.25. \tag{128}$$

When interpreting the observed cross-correlation skewness it is important to bear in mind that different factors can be responsible for skewing: random velocity spread in the "wind" pattern; variation of the inhomogeneities in the "vorticity" pattern; refraction of radio waves at large-scale inhomogeneities; effect of internal noise of the detection equipment.

Let us inquire how refraction at large-scale inhomogeneities affects the rearrangement of the IPS pattern. The refraction angle is

$$\theta_r = \frac{1}{k} \nabla_\perp \Phi, \tag{129}$$

where k is the wave number and Φ is the phase lead of the wave in the medium. Then for the linear ray displacement corresponding to refraction angle θ_r we have

$$\Delta x = z\theta_r, \tag{130}$$

where z is the distance from the observer to the inhomogeneities. The variable pattern velocity **u** associated with refraction is

$$\mathbf{u} = \frac{d\Delta\mathbf{x}}{dt} = \frac{z}{k}\frac{d}{dt}\nabla_\perp[\Phi(\mathbf{r} - \mathbf{V}\,t)] = \frac{z}{k}\mathbf{V}\frac{\partial}{\partial\mathbf{r}}\nabla_\perp\Phi(\mathbf{r}) \sim \frac{z\theta_\mathbf{r}}{L}\mathbf{V}, \qquad (131)$$

where L is the space scale of the inhomogeneities responsible for refraction and **V** is the velocity of the inhomogeneities. We estimate the velocity $|\mathbf{u}|$, specifying the parameters of the interplanetary plasma. We know that the space scale of the large inhomogeneities $L \sim 10^{11}$ cm [68-70], the solar-wind velocity $V \sim 4 \cdot 10^7$ cm/sec, and the phase lead at large inhomogeneities $\Phi \sim 10^4$ rad [46]. For these parameters of the interplanetary plasma we obtain

$$\theta_\mathbf{r} \sim 10^{-6}, \quad u \sim 10^{-4}\,V. \qquad (132)$$

Estimate (132) shows that the expected velocity spread associated with refraction is negligible in comparison with the observed spread $\sigma/\langle V\rangle \simeq 0.25$. Also, the characteristic time of variation of the velocity in the refraction effect is of order $T_\mathbf{r} \sim 10^4$ sec, which is greater than the observation period. Consequently, the observed cross-correlation skewness and rearrangement of the IPS pattern cannot be attributed to refraction at large-scale inhomogeneities.

We now consider the influence of internal noise of the detection equipment on the cross-correlation skewness. The noise problem has been investigated in [71], in which it is shown that the mean-square τ-asymmetric noise increment to the cross-correlation function $\rho(\mathbf{r}, \tau)$ is

$$\langle\Delta\rho^2\rangle = 2\frac{\tau_s}{T}[m^2(\Delta s_1^2 + \Delta s_2^2)], \qquad (133)$$

where τ_s is the equipment time constant, m is the scintillation index, $\Delta s_{1,2}^2$ denotes the variances of uncorrelated noise at the observation points, and T is the observation period. Using (133) in the definition of the skewness parameter (34), we find the noise skewness

$$S^{noi} \simeq 2\,(\tau_s/T)^{1/2}\,(\Delta s^2/m^2)^{1/2}. \qquad (134)$$

Substituting into (134) the values $\tau_s \sim 0.1$ sec, $T \sim 10^3$ sec, and $\Delta s/m \sim 0.1$, we obtain $S^{noi} \sim 10^{-3}$, which is an order of magnitude smaller than the observed skewness S^{obs}. The estimate $S^{noi} \sim 10^{-3}$ refers to the rms value of the noise skewness. In individual cases the noise can be rather high ($S^{noi} \sim S^{obs}$), but the average value of S^{noi} is equal to zero ($\langle S^{noi}\rangle = 0$). Therefore, to eliminate the influence of noise it is necessary to use a value of S averaged over a certain number of records, as is done in [39].

Thus, to explain the observed cross-correlation skewness only two possibilities are left: actual velocity spread ("wind" pattern) and time variation of the inhomogeneities ("vorticity" pattern). Cross-correlation analysis does not permit differentiation between these factors. Consequently, for a conclusive explanation of the nature of the observed velocity fine structure we must recruit data obtained from space vehicles on the velocities of the interplanetary medium.

Armstrong and Coles [18] have analyzed the drift of the scintillation pattern of the source 3C144 at frequency $f = 74$ MHz (1971) from three-station observations with baselines of 94, 94, and 71 km (San Diego, California). The analysis was carried out by the BPS, LE, and intersection methods, which, as mentioned in Section 1, yield consistently low values for the average drift velocity. Accordingly, the SD of the velocities of the inhomogeneities was estimated by indirect means, by comparing the apparent velocities with the velocities of the proton component of the solar wind, measured on board Pioneers 6 and 7. Allowance was made for the extent of the medium on the apparent velocity of the IPS pattern. The values obtained for the velocity SD amount to 25-30% of the average velocity $\langle V \rangle \simeq 450$ km/sec.

Coles and Maagoe [72] have determined the velocity spread ΔV as follows (San Diego): A certain correlation level was drawn for the function $\rho(\mathbf{r}, \tau)$ (usually 1/2), and then the time $\tau_3 = \tau_0 + \frac{1}{2}(t_2 - t_1)$ was computed, where τ_0 is the position of the maximum of $\rho(\mathbf{r}, \tau)$, $\rho(\mathbf{r}, \tau_0 + t_2) = \rho(\mathbf{r}, \tau_0 - t_1) = \frac{1}{2}$ (see Fig. 2). The time τ_3 was used to determine the velocity $V_3 = r/\tau_3$ (which is close to the average velocity of the inhomogeneities). The velocity spread was determined as the difference between the apparent velocity V_a and the velocity V_3:

$$\Delta V = V_a - V_3 \tag{135}$$

It is seen at once that the definition of ΔV (135) is based on the presence of skewness in the function $\rho(\mathbf{r}, \tau)$. Thus, the time τ_3 can be written as follows on the basis of the definition (47) for the skewness S:

$$\tau_3 = \tau_0 + \frac{1}{2}(t_2 - t_1) = \tau_0 + \frac{1}{2} S (t_2 + t_1) \simeq \tau_0 (1 + S). \tag{136}$$

Now from expressions (135) and (136) we obtain for ΔV

$$\Delta V = \frac{r}{\tau_0} - \frac{r}{\tau_0 (1 + S)} \simeq \frac{r}{\tau_0} S = V_a \frac{r}{a} \frac{\sigma^2}{\langle V \rangle^2 + \sigma^2} = \frac{r}{a} \frac{\sigma^2}{\langle V \rangle^2}, \tag{137}$$

whence we arrive at the velocity SD

$$\sigma = \left(\frac{a}{r} \langle V \rangle \Delta V \right)^{1/2} \simeq \left(\frac{a}{r} V_a \Delta V \right)^{1/2}. \tag{138}$$

The estimate obtained for ΔV in [72] is $\Delta V \lesssim 50$ km/sec. Substituting $\Delta V \lesssim 50$ km/sec, $a \simeq 100$ km, $r \simeq 90$ km, and $\langle V \rangle \simeq 400$ km/sec into (138), we obtain an estimate for σ:

$$\sigma \lesssim 150 \text{ km/sec}, \tag{139}$$

which is not inconsistent with estimate (128) or the SD σ obtained in [18]. We note that Chashei and Shishov [71] have observed the interesting effect of an increase in the velocity SD σ with the heliolatitude. In the model of the medium with velocity structure the effective thickness of the scattering region must increase with the heliolatitude, so the velocity spread increases as well. Consequently, the tendency for the velocity spread to increase with the heliolatitude, as observed experimentally, is consistent with the hypothesis of the wind nature of the scintillation pattern.

The most complete data on the cross-correlation analysis of IPS drift are published in the survey [12], in which particular attention is given to the behavior of the velocity SD as a function of the elongation. The velocity SD σ in the interval of elongations $\rho > \frac{1}{2}$ is $\sigma \simeq 100$ km/sec and is practically independent of the elongation. As the line of slight approaches the position $\varepsilon = 90°$ ($\rho = 1$ a.u.) a certain increase is noticed in the velocity SD. This result is in good agreement with the results of the papers cited above.

Thus, the experimental data obtained by the method of cross-correlation analysis disclose the presence of velocity fine structure in the interplanetary medium. The standard deviation of the velocities σ amounts to 25-30% of the average velocity of the inhomogeneities. The numerical value of the velocity SD and its behavior as a function of the elongation and heliolatitude are not inconsistent with the model of the medium in which distortions of the form of the cross-correlation function are associated with the presence of physically different velocities along the line of sight. The disparity between the solar-wind velocity measured on rockets and the apparent pattern velocity ($V_a \simeq 400$ km/sec) is slight and corresponds to the 25-30% velocity SD. The BPS, LE, and intersection methods yield far too low a value for the velocity of the inhomogeneities, so that the most realistic quantity is the apparent velocity of the diffraction pattern, and the velocity SD must be determined with allowance for cross-correlation skewness.

3.2. Dispersion Analysis

We now discuss the dispersion functions $V(\omega)$ obtained from IPS observations at several stations, using the dispersion-analysis theory developed in Section 2.

Golley and Denisson [39] have obtained the dispersion functions $V(\omega)$ for the radio source 3C48, which was observed in 1966 from three stations with baselines of 85, 66, and 52 km, at a frequency of 81.5 MHz. The individual functions $V(\omega)$ are generally increasing functions, consistent with the results of the theoretical discussion in Section 2. An analysis of the individual functions $V(\omega)$ leads to the important conclusion that rather high velocities are essentially always present in the diffraction pattern ($V > 500$ km/sec). This conclusion cannot be deduced within the context of cross-correlation analysis and indicates that the model of a spherically symmetric solar wind cannot fully account for the observed velocity SD. The velocity spread in $\langle V(\omega) \rangle$, averaged over 14 samples (Fig. 8), is $\simeq 250$ km/sec for an average velocity $\langle V \rangle \simeq 400$ km/sec [4]. As shown in Section 2, the velocity SD is always less than the velocity spread in the dispersion function. From the function $\langle V(\omega) \rangle$ (Fig. 8) we can estimate the velocity SD σ. For the model of the medium with a normal velocity distribution and a Gaussian spectrum of inhomogeneities with scale $a \simeq 160/\sqrt{2}$ km [4] the best fit to the experimental function $\langle V(\omega) \rangle$ is given by an analytical function $V(\omega)$ with average velocity $\langle V \rangle = 410$ km/sec and a relative velocity SD $\sigma/\langle V \rangle = 0.35$.

The authors have plotted the dispersion functions $V(\omega)$ according to observations of the scintillations of the sources 3C144 and 3C48 in May, 1975, in San Diego at a frequency of 74 MHz (three stations with baselines r $\simeq 90$ km). The period of observation of the source in San Diego was T > 1 h, which is an order of magnitude greater than the observation period in Cambridge [39]. Consequently, the noise content of the individual functions $V(\omega)$ obtained from the San Diego data is much lower than that of $V(\omega)$ according to the Cambridge data [39]. There are two types of observed individual functions $V(\omega)$: monotonically growing and arriving at a constant level. The growth of $V(\omega)$ is consistent with the theoretical analysis in Section 2. If $V(\omega)$ arrives at a constant level $V(\omega) = V_{max}$, then this is evidence that there are no inhomogeneities with velocities greater than V_{max} along the line of sight. If $V(\omega)$ does not settle into a plateau in the interval $\omega < \omega_{max}$ (ω_{max} being determined by the noise level), then this means that either inhomogeneities with velocities $V \geq V(\omega_{max})$ are present along the line of sight or the inhomogeneity space scales in high-velocity regions of the interplanetary plasma are smaller than the inhomogeneity scales in low-velocity regions. The functions $V(\omega)$ change from day to day. This fact indicates a change in the velocity structure along the line of sight, possibly due to nonsteadiness of the flow of matter from the sun or to co-rotation of large-scale formations of the interplanetary plasma.

The functions $\langle V(\omega) \rangle$ averaged over five days of observations for the sources 3C144 and 3C48 are given in Fig. 9. In averaging for each observation session, from three functions

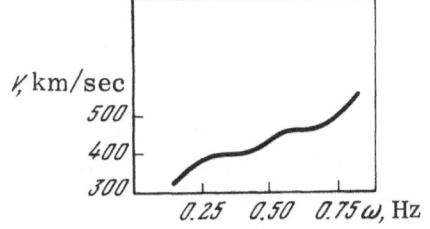

Fig. 8. Behavior of $\langle V(\omega) \rangle$ for the radio source 3C48 [39].

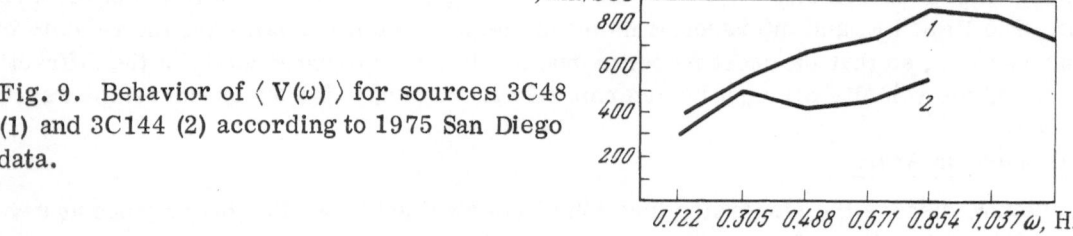

Fig. 9. Behavior of $\langle V(\omega) \rangle$ for sources 3C48 (1) and 3C144 (2) according to 1975 San Diego data.

$V(\omega)$ the one was chosen with minimum angle between the velocity and the baseline. A comparison of the functions $\langle V(\omega) \rangle$ in Fig. 9 with the theoretical functions for the model of an extended medium (see Section 2.2) shows that the model of a spherically symmetric solar wind with $\Delta n^2(\rho) \propto \rho^{-4}$ does not describe the observed functions $\langle V(\omega) \rangle$. Therefore, the velocity distribution $\varphi(V)$ in the solar wind is flatter then the distribution function (42) for a spherically symmetric wind. Fitting various distributions $\varphi(V)$ to the observed functions $V(\omega)$ (Fig. 9) yields the following bounds for the relative velocity SD:

$$0.25 < \sigma/\langle V \rangle < 0.34. \tag{140}$$

These bounds are not inconsistent with the bounds obtained for $\sigma/\langle V \rangle$ within the framework of cross-correlation analysis and according to $\langle V(\omega) \rangle$ from [39] (see Fig. 8).

A comparison of the functions $\langle V(\omega) \rangle$ for the sources 3C144 (low heliolatitudes) and 3C48 (high heliolatitudes) shows that the average velocity of the inhomogeneities at high heliolatitudes is greater than at low heliolatitudes. Moreover, it is evident from Fig. 9 that the velocity spread for 3C48 is greater than for 3C144, although the relative velocity SDs $\sigma/\langle V \rangle$ for these sources are roughly the same.

We now attempt to answer the question of the nature of the observed velocity spread, i.e., the nature of the scintillation pattern, whether it is wind or vorticity. We know from spaceship measurements of the solar-wind velocity that the latter varies within the limits $\simeq 300$ to $\simeq 800$ km/sec at a fixed point in space, with a characteristic time scale of order 1 day [73]. The velocity SD σ_I associated with this effect is $\sigma_I \simeq 20\%$ of the average velocity $\langle V \rangle \simeq 450$ km/sec. These variations can be attributed either to a large-scale layered structure of the flow of matter from the sun or with nonsteady flow in the spherically symmetric model. The effective scattering thickness for radio waves in the interplanetary medium at a distance $\rho \simeq 0.5$ a.u. from the sun corresponds to a time scale of ~ 5 days. It may be stated, therefore, that the velocity variations recorded on rockets will create a velocity structure along the line of sight and be manifested in the scintillation phenomenon. The effect of velocity inhomogeneity along the line of sight is superposed on the geometric effect associated with variation of the velocity projection, the velocity SDs from these effects being added independently on the average. The total velocity SD associated with the geometric effect $\sigma_G = 0.2\langle V \rangle$ (43) and with inhomogeneity of the velocity along the line of sight $\sigma_I = 0.2\langle V \rangle$ is now equal to

$$\sigma = (\sigma_G^2 + \sigma_I^2)^{1/2} \simeq 0.28 \langle V \rangle. \tag{141}$$

Estimate (141) is highly consistent with the relative velocity SD according to cross-correlation analysis (128) and according to dispersion analysis (140).

Thus, the experimental data analyzed by the cross-correlation and dispersion methods afford a direct assessment of the velocity-SD bounds: $0.25 < \sigma/\langle V \rangle < 0.35$, which agree with the bounds deduced indirectly in [18]. It follows from the latter that the velocity SD in the IPS is $\sigma \lesssim 150$ km sec. The main contribution to the observed velocity SD is from the geometric effect in conjunction with inhomogeneity of the instantaneous velocity distribution along the line of sight. This means that the rearrangement of the scintillation pattern can be attributed to the presence of velocity find structure, and the hypothesis of the wind nature of the scintillation pattern does not contradict the experimental data. The structure of the velocities along the line of sight varies from one day to the next and with the heliolatitude. The conclusion of the wind nature of the scintillation pattern is applicable to regions comparatively far from the sun ($\rho \gtrsim 0.5$ a.u.) and affords a lower bound for the lifetime of the inhomogeneities τ_*, which must be greater than the time for wind rearrangement of the pattern τ_c:

$$\tau_* > \tau_c = a/\sigma = \tau_1 \langle V \rangle / \sigma \simeq 3 \text{ sec.} \tag{142}$$

There is evidence that the velocity SD increases as the sun is approached [38]. This effect cannot be explained in the model of a spherically symmetric solar wind. One possible explanation lies in the fact that the inhomogeneity lifetime τ_* decreases drastically in regions close to the sun and inequality (142) fails. This event can be caused by growth of the phase velocities and natural frequencies of plasma turbulence (see Section 2.2d). If the second hypothesis is true, then it should be stated that inhomogeneities in the solar wind are generated by a local source (i.e., as a result of instability). Accordingly, it would be most interesting to conduct a detailed study of the velocity fine structure at small elongations and, for that purpose, to use the methods proposed in the present article, dispersion analysis in particular.

LITERATURE CITED

1. V. L. Ginzburg, Dokl. Akad. Nauk SSSR, 109:61 (1956).
2. A. Hewish, P. F. Scott, and D. Wills, Nature, 203:1214 (1964).
3. T. D. Antonova, V. V. Vitkevich, and V. I. Vlasov, Tr. FIAN, 38:88 (1967).
4. P. A. Denisson and A. Hewish, Nature, 213:343 (1967).
5. M. H. Cohen, Astrophys. J., 147:449 (1967).
6. R. V. E. Lovelace, E. E. Salpeter, and L. Sharp, Astrophys. J., 159:1047 (1970).
7. G. Bourgous, Astron. Astrophys., 19:200 (1972).
8. T. Watanabe and T. Kakinuma, Publ. Astron. Soc. Jpn., 24:459 (1972).
9. A. Pramesh Rao, S. M. Bhandari, and S. Ananthakrishanan, Aust. J. Phys., 27:105 (1974).
10. Z. Houminer, Planet. Space Sci., 21:1367 (1973).
11. T. D. Shishova, Astron. Tsirk., No. 819 (1974).
12. W. A. Coles, B. J. Rickett, and V. H. Rumsey, in: Solar Wind III (C. T. Russel, ed.), Univ. California, Los Angeles (1974), p. 351.
13. B. J. Rickett, Monthly Not. Roy. Astron. Soc., 150:67 (1970).
14. G. G. Getmantsev and L. M. Eroukhimov, Ann. IQSY, 5:229 (1969).
15. C. L. Rufenach, J. Geophys. Res., 79:1562 (1974).
16. W. M. Cronyn, Astrophys. J., 161:760 (1970).
17. I. V. Chashei, Astron. Zh., 52:365 (1975).
18. J. W. Armstrong and W. A. Coles, J. Geophys. Res., 77:4602 (1972).
19. B. H. Briggs, G. J. Philips, and D. H. Shinn, Proc. R. Soc. London, Ser. B, 63:106 (1950).
20. N. A. Lotova and I. V. Chashei, Astron. Zh., 50:348 (1973).
21. I. A. Alekseev, V. V. Vitkevich, V. I. Vlasov, Yu. P. Ilyasov, S. M. Kutuzov, and M. M. Tyaptin, Tr. FIAN, 47:183 (1969).
22. S. F. Mirkotan and Yu. V. Kushnerevskii, Ionospheric Research [in Russian], No. 12, Nauka, Moscow (1964).
23. V. D. Gusev and S. F. Mirkotan, Vestn. Mosk. Univ., Ser. Fiz.-Mat., No. 3, p. 4 (1961).
24. L. S. Fedor, J. Geophys. Res., 72:5401 (1972).
25. R. D. Ekers and L. T. Little, Astron. Astrophys., 10:310 (1971).
26. V. I. Shishov, Zh. Éksp. Teor. Fiz., 61:1399 (1971).

27.   J. A. Fejer, Proc. R. Soc. London, Ser. A, 220:455 (1953).

28.   V. V. Pisareva, Astron. Zh., 35:112 (1958).

29.   R. F. Mercier, Proc. Cambridge Philos. Soc., Ser. A, 58:382 (1962).

30.   N. A. Lotova, Usp. Fiz. Nauk, 95:293 (1968).

31.   E. E. Salpeter, Astrophys. J., 147:443 (1967).

32.   J. T. Goslin et al., J. Geophys. Res., 77:5442 (1972).

33.   Z. Houminer, Planet. Space Sci., 21:1617 (1973).

34.   N. A. Lotova and I. V. Chashei, Geomagn. Aéron., 15:193 (1975).

35.   N. A. Lotova and I. V. Chashei [Chashey], Astrophys. Space Sci., 20:251 (1973).

36.   N. A. Lotova and I. V. Chashei, Izv. Vyssh. Uchebn. Zaved., Radiofiz., 16:491 (1973).

37.   N. A. Lotova and I. V. Chashei, Radiotekh. Élektron., 20:1777 (1975).

38.   L. T. Little and R. D. Ekers, Astron. Astrophys., 10:306 (1971).

39.   M. G. Golley and P. A. Denisson, Planet. Space Sci., 18:95 (1970).

40.   C. R. McGee, J. Atmos. Terrest. Phys. 28:861 (1966).

41.   V. I. Shishov, Astron. Zh., 49:1258 (1972).

42.   N. A. Lotova and I. V. Chashei, Geomagn. Aéron., 13:998 (1973).

43.   N. A. Lotova and I. V. Chashei, Geomagn. Aéron., 15:769 (1975).

44.   B. H. Briggs, J. Atmos. Terrest. Phys., 30:1777 (1968).

45.   B. H. Briggs and M. G. Golley, IAGA-URSI Symp., St. Gallen, Switzerland (1967).

46.   J. R. Jokipii and J. V. Hollveg, Astrophys. J., 160:745 (1970).

47.   A. Hewish, Astrophys. J., 163:645 (1971).

48.   B. J. Rickett, J. Geophys. Res., 78:1543 (1973).

49.   N. A. Lotova and I. V. Chashei, Geomagn. Aéron., 12:800 (1972).

50.   W. A. Coles and J. K. Harmon, Preprint Univ. California 92037 (1974).

51.   A. T. Young, Astrophys. J., 168:543 (1971).

52.   J. R. Jokipii and L. C. Lee, Astrophys. J., 172:729 (1972).

53.   J. R. Jokipii and L. C. Lee, Astrophys. J., 182:317 (1973).

54.   V. I. Tatarskii, Wave Propagation in a Turbulent Atmosphere [in Russian], Nauka,
      Moscow (1967).

55.   N. A. Lotova and A. A. Rukhadze, Astron. Zh., 45:343 (1968).

56.   N. A. Lotova and I. S. Baikov, Astron. Zh., 46:1057 (1969).

57.   D. W. Forslund, J. Geophys. Res., 75:17 (1970).

58.   F. Perkins, Astrophys. J., 179:637 (1973).

59.   N. A. Lotova, Usp. Fiz. Nauk, 115:603 (1975).

60.   V. N. Tsytovich, Theory of Turbulent Plasmas [in Russian], Atomizdat, Moscow (1971).

61.   M. G. Morgan and K. L. Bowles, Science, 161:1139 (1968).

62.   V. V. Vitkevich and V. I. Vlasov, Astron. Tsirk., No. 396 (1966).

63.   A. Hewish and M. D. Symonds, Planet. Space Sci., 17:313 (1969).

64.   J. W. Armstrong. W. A. Coles, and J. K. Harmon, Am. Geophys. Union Fall Annu. Meet-
      ing, Paper SS3, San Francisco (1973).

65.   T. Watanabe, K. Shibasaki, and T. Kakinuma, J. Geophys. Res., 79:3841 (1874).

66.   V. V. Vitkevich and V. I. Vlasov, Astron. Zh., 49:595 (1972).

67.   N. A. Lotova and N. V. Vereshchagina, Izv. Vyssh. Uchebn. Zaved., Radiofiz., 16:1645
      (1973).

68.   J. R. Jokipii and P. J. Colemann, J. Geophys. Res., 73:5495 (1968).

69.   B. E. Goldstein and G. I. Siscoe, Solar Wind (C. P. Sonett, P. J. Colemann, and Y. Wilcox,
      eds.), NASA, Washington, DC (1972).

70.   D. S. Intrilligator and J. H. Wolf, Astrophys. J., 162:187 (1970).

71.   I. V. Chashei and V. I. Shishov, Pis'ma Astron. Zh., 1:18 (1975).

72.   W. A. Coles and S. Maagoe, J. Geophys. Res., 77:5622 (1972).

73.   P. J. Colemann, Astrophys. J., 153:371 (1968).

INFLUENCE OF THE DOPPLER EFFECT IN MEASUREMENTS OF THE PERIODS OF PULSARS

T. V. Shabanova

The influence of observer motion and the concomitant Doppler effect on the measured value of a pulsar period is analyzed. A method is given for computing the corrections needed for adjustment of the pulse observations to a barycentric coordinate system.

INTRODUCTION

Pulsars emit strictly periodic pulses, the periods of which remain stable within error limits of 10^{-15} sec per period. This extraordinarily high periodicity of pulse emission demands the consideration of every type of observer motion that can affect the measured value of the period, i.e., the observations must be referred to a fixed barycentric coordinate system. Corrections for observer motion can be introduced at the times when pulses are observed. These corrections are discussed in the first part of the article. The corrections introduced in the observed periods of pulsars are calculated in the second part.

1. Reduction of Pulse Observation Times to the Barycenter of the Solar System

We first reduce the observations to a geocentric coordinate system. An observer situated at a particular point on the earth's surface receives pulses at times differing from those at which the pulses arrive at the center of the earth. This difference depends on the position of the observer relative to the meridian. It decreases with distance of the observer from the meridian and is equal to zero for an hour angle t = 90°, i.e., at this point the pulses arrive simultaneously at the observer and at the center of the earth. To compute this correction it is necessary to project the radius of the earth at the observation point onto the line joining the center of the earth with the pulsar and to calculate the time required for a light wave to traverse this distance. For an observer at latitude φ (Fig. 1) the projection of the earth's radius is $r = r_0 \cos(\varphi - \delta_\Sigma) \cdot \cos t$, and the time required to traverse this distance is

$$\Delta T_1 = \frac{r_0}{c} \cos(\varphi - \delta_\Sigma) \cdot \cos t,$$

where r_0 is the mean equatorial radius of the earth, r_0 = 6 378 160 m, c is the speed of light, φ is the latitude of the observer, and δ_Σ, t are the inclination and hour angle of the pulsar. Consequently, in transformation from a topocentric to a geocentric coordinate system it is necessary to introduce the following correction at the observation times:

$$\Delta T_1 = 0.021\ 275\ \cos(\varphi - \delta_\Sigma) \cdot \cos t,$$

where ΔT_1 is expressed in seconds.

125

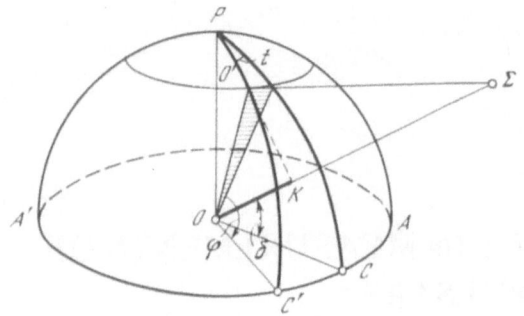

Fig. 1. Transformation from a topocentric to a geocentric coordinate system.

Fig. 2. Relationship between geocentric and barycentric coordinate systems. QQ', ecliptic plane, Π, ecliptic pole; T, position of earth in ecliptic plane; B, barycenter of solar system; AA', plane of equator; P, pole of equator.

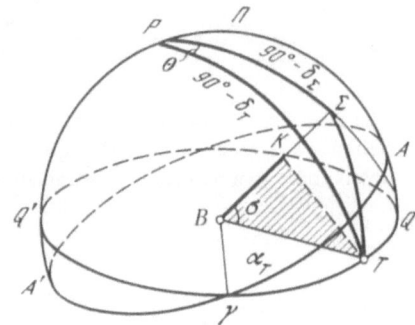

To eliminate the influence of the earth's motion in its heliocentric orbit and the associated Doppler effect due to the nonuniformity of arrival of pulsar pulses at the observer during the year, the pulse observation times must be reduced to a barycentric coordinate system. The correction here is computed by a procedure analogous with the preceding. The projection of the radius of the earth's orbit onto the barycenter–pulsar line (Fig. 2) is determined, and the time for a light wave to traverse this distance is calculated. The projection of the earth's radius vector R_0 onto the barycenter–pulsar line is $R = R_0 \cos \sigma$, where the angle σ is evaluated from the spherical triangle $P\Sigma T$ and is equal to $\cos \sigma = \sin \delta_T \sin \delta_\Sigma + \cos \delta_T \cdot \cos (\alpha_\Sigma - \alpha_T)$. To find the earth's radius vector R_0, along with the direct ascension and inclination of the earth (α_T, δ_T) we use the rectangular equatorial barycentric coordinates of earth X, Y, Z:

$$X = R \cos \delta_T \cdot \cos \alpha_T,$$

$$Y = R \cos \delta_T \cdot \sin \alpha_T,$$

$$Z = R \sin \delta_T,$$

which, in turn, are found from the relations [1]

$$X = -\left[X_\odot + \frac{\sum m_i x_i}{1 + \sum m_i} \right],$$

$$Y = -\left[Y_\odot + \frac{\sum m_i y_i}{1 + \sum m_i} \right],$$

$$Z = -\left[Z_\odot + \frac{\sum m_i z_i}{1 + \sum m_i} \right],$$

where $X_\odot$, $Y_\odot$, $Z_\odot$ are the rectangular equatorial geocentric coordinates of the sun, and x_i, y_i, z_i are the rectangular equatorial heliocentric coordinates of the i-th planet, whose mass m_i is expressed in units of the mass of the sun. The rectangular coordinates of the sun and the planets are published in the Astronomical Yearbook of the USSR, and the masses of the planets may be found in the International Astronomical Union (IAU) System of Astronomical Constants.

The transformation to barycentric coordinates is realized by introducing the following corrections at the observation times:

$$\Delta T_2 = 499.012 \ R \ [\sin \delta_T \cdot \sin \delta_\Sigma + \cos \delta_T \cdot \cos \delta_\Sigma \cdot \cos (\alpha_\Sigma - \alpha_T)],$$

where α_Σ, δ_Σ are the average coordinates of the pulsar and ΔT_2 is expressed in seconds.

The correction ΔT_2 has value zero for pulsars situated at the ecliptic pole and is a maximum for pulsars in the ecliptic plane, varying from zero to 499 sec in the course of the year. The principal error in the computation of ΔT_2 is incurred by imprecise knowledge of the pulsar coordinates.

The motion of the earth in the variable gravitational field of the sun, according to general relativity theory, will cause the slowing of clocks and variations of atomic-clock time (ACT) in the course of the year. The variation of ACT during the year depends on the earth's position in orbit and is described by the equation [2]

$$k = t - s = 0.001 \ 661 \ 45 \ [\sin l + \tfrac{1}{2} \sin 2l + \tfrac{3}{8} e^2 (\sin 3l - 3 \sin l)],$$

in which t is coordinate-time, or the time on rest clocks outside the field of gravity; s is local time, measured by atomic clocks on the moving earth; l is the average anomaly of earth; and e is the eccentricity of earth's orbit. According to this equation, the pulsar period must be a minimum in January, when the earth is at the perihelion, and a maximum in June, at the aphelion, or the difference between the measured time of the pulse observations and the calculated time will be negative in April ($s - t = -0.00166$ sec) and positive in October ($s - t = 0.00166$ sec), since atomic clocks run slower in the first half of the year than in the second half.

Thus, to obtain the pulse times in the barycentric coordinate system it is necessary to intoduce the following corrections at the pulse observation times t_0:

$$t_B = t_0 + \Delta T_1 + \Delta T_2 + k.$$

2. Calculation of Corrections to Observed Period

The periods of the majority of pulsars are measured with high precision, and their values are given in the barycentric coordinate system. However, situations frequently arise in which the period must be known in a moving coordinate system, for example, in observations of pulsars in the synchronous-buildup regime. In transformation from the barycentric to a topocentric coordinate system (and vice versa) it is necessary to take into account the influence of all kinds of observer motion on the value of the period.

Following are the types of motion involving the observer: (1) rotation of the earth about its own axis; (2) motion of the earth's center of gravity (COG) about the COG of the earth−moon system; (3) motion of the COG of the earth−moon system about the sun; (4) motion of the COG of the sun about the COG of the solar system.

It is known from basic physics that in the case of observer motion the pulse-emission period varies according to the Doppler equation

$$P_0 = P_0 \ (1 - V \cos \sigma/c),$$

in which $V \cos \sigma$ is the projection of the observer velocity onto the line of sight, P_0 and P_0 are the true and observed pulsar periods, and c is the velocity of light. The observer velocity V is considered to be positive for motion toward, and negative for motion away from the emission source.

We now examine in succession the influence of each of the above-stated types of observer motion on the pulsar period. In each case we determine the projection $V \cos \sigma$ of the observer velocity onto the line of sight. We denote the dimensionless quantity $V \cos \sigma / c$ by ΔA.

1. **Correction for Axial Rotation of the Earth.** In observations of a pulsar the detected period will be somewhat shorter before its culmination than at the instant of culmination and larger after culmination. To find the projection of the observer velocity onto the direction $O\Sigma$ we examine Fig. 3, in which P is the pole of the equator, Z is the zenith, O is the observer, Σ is the pulsar, and t is the hour angle of the pulsar. From the spherical triangle $WP\Sigma$, in which the side $PW = 90°$, $P\Sigma = 90° - \delta_\Sigma$, $W\Sigma = 180° - \sigma$, and the angle $\theta = 90° - t$, we have according to the cosine formula $-\cos \sigma = \cos \delta_\Sigma \cdot \sin t$. The projection of the observer velocity in the direction $O\Sigma$ is

$$V \cos \sigma = -V_0 \cos \varphi \cdot \cos \delta_\Sigma \cdot \sin t,$$

where $V_0 \cos \varphi$ is the linear velocity of the observer at latitude φ and V_0 is the equatorial velocity, equal to $V_0 = 2\pi r_0 / 86\ 164 = 0.4651$ km/sec. The required correction has the form

$$\Delta A_1 = -0.1551 \cdot 10^{-5}\ \cos \varphi \cdot \cos \delta_\Sigma \cdot \sin t.$$

2. **Correction for Motion of the Earth−Moon COG.** It is assumed that during one month the earth's COG executes uniform circular motion in the ecliptic plane about the COG of the earth−moon system with linear velocity V_0. During this same period the observed pulsar period varies sinusoidally with amplitude $V_0 \cos \beta_\Sigma$. Let the moon be situated at a point with geocentric longitude l, so that its velocity has longitude $(l + 90°)$ and the velocity

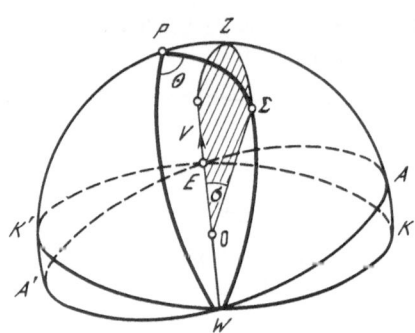

Fig. 4. Motion of the earth about COG of the earth−moon system.

Fig. 3. Axial rotation of the earth.

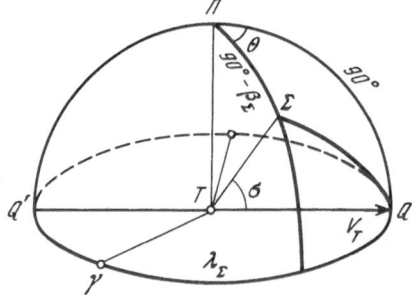

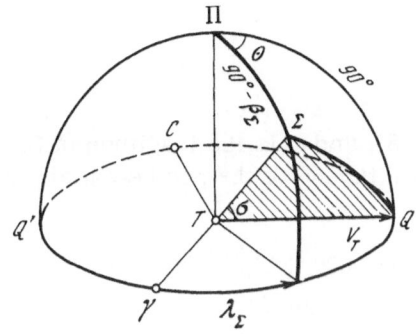

Fig. 5. Orbital rotation of the earth about the
sun.

of earth is directed toward a point with longitude $(l + 90° + 180°)$. The projection of the earth's
velocity onto the earth−pulsar line is (Fig. 4)

$$V \cos \sigma = V_0 \cos \beta_\Sigma \cdot \sin (l - \lambda_\Sigma),$$

where β_Σ, λ_Σ are the ecliptic latitude and longitude of the pulsar and V_0 is the linear velocity
of the earth, equal to [3]

$$V_0 = \frac{2\pi R_{TM}}{T} \frac{m_M}{m_M + m_T} = 0.0124 \text{ km/sec.}$$

We finally obtain

$$\Delta A_2 = 0.4133 \cdot 10^{-7} \cos \beta_\Sigma \cdot \sin (l - \lambda_\Sigma).$$

3. Correlation for Orbital Motion of the Earth about the Sun.
The main factor contributing to the variation of the observed pulsar period is motion of the
earth about the sun. To obtain the projection of the earth's orbital velocity onto the earth−
pulsar line we examine Fig. 5, which shows the celestial sphere with its center at the center
of the earth; Π is the pole of the ecliptic, γ is the point of the vernal equinox, and S, Σ are the
positions of the sun and the pulsar. Due to the ellipticity of the earth's orbit the angle between
the direction of the earth's velocity TQ and the earth−sun line TS is equal to $(90° + i)$, and not
to 90° as in the case of circular motion. The required angle σ is determined from the spheri-
cal triangle $\Pi\Sigma Q$, in which the arcs $\Pi Q = 90°$, $\Pi\Sigma = 90° - \beta_\Sigma$, and the angle at the ecliptic pole is $\theta =$
$\lambda_\odot - (90° + i) - \lambda_\Sigma$. Hence,

$$\cos \sigma = \sin (90° - \beta_\Sigma) \cdot \cos (\lambda_\odot - 90° - i - \lambda_\Sigma) = \cos \beta_\Sigma \cdot \sin (\lambda_\odot - \lambda_\Sigma - i),$$

and the projection of the earth's line of sight is equal to

$$V \cos \sigma = V_0 \cos \beta_\Sigma \cdot \sin (\lambda_\odot - \lambda_\Sigma - i), \tag{1}$$

The velocity V_0 of the earth in orbit is a variable and in the theory of elliptical motion
is given by the expression [3, 4]

$$V_0 = \frac{2\pi}{T} \frac{a}{\sqrt{1 - e^2}} [1 + e \cos (\lambda_\odot - v)] \frac{1}{\cos i},$$

in which T is the duration of the sidereal year, equal to 366.2421988×86164 sec, where 86164
is the number of seconds mean time in sidereal days, and a, e denote the semimajor axis and

eccentricity of the earth's orbit. The expression for the angle i has the form

$$\tan i = \frac{e \sin (\lambda_\odot - v)}{1 + e \cos (\lambda_\odot - v)} \; ;$$

here v is the longitude of the perigee, $v = \omega + 180° \approx 282.5°$, and ω is the longitude of the perihelion. The eccentricity e of the earth's orbit and the longitude ω of the perihelion are variable and are given by the expression [4]

$$e = 0.016\ 751\ 04 - 0.000\ 041\ 80 t - 0.000\ 000\ 126 t^2,$$
$$\omega = 101°13'15.0'' + 6189.03'' t + 1.63'' t^2 + 0.012'' t^3.$$

The time t is reckoned in Julian centuries as 36525 ephemeris days from the initial epoch 1900, Jan. 0.12 E.T. = JD 2415020.0 until the time in question t_0:

$$t = (JD\ (t_0) - 2{,}415{,}020.0)/36{,}525.$$

During one year the orbital velocity varies from 29.2 to 30.3 km/sec, with an average value of 29.76 km/sec, and the angle i varies from +56' to −56'. Substituting the expressions for the orbital velocity V_0 and angle i into Eq. (1) and bearing in mind that the ratio a/c is the light-astronomical unit, equal to 499.012 sec, we obtain the following expression for the correction due to orbital motion of the earth:

$$\Delta A_3 = 499.012 \frac{2\pi}{T} \frac{[1 + e \cos (\lambda_\odot - v)]}{\sqrt{1 - e^2}} \frac{1}{\cos i} \cos \beta_\Sigma \cdot \sin (\lambda_\odot - \lambda_\Sigma - i).$$

The influence of the correction ΔA_3 on the pulsar period is significant and can alter it by as much as several fractions of a millisecond. Figure 6 gives a curve of the annual variation of the period for the pulsar PSR 1133 + 16. The rate of change of ΔA_3 is fairly high:

$$\frac{d}{dt}(\Delta A_3) = \frac{2\pi}{T} \frac{a}{c} \frac{[1 + e \cos (\lambda_\odot - v)]}{\sqrt{1 - e^2}} \frac{1}{\cos i} \cos \beta_\Sigma \cdot \cos (\lambda_\odot - \lambda_\Sigma - i) \simeq 10^{-4} \cos \beta_\Sigma \cdot \cos (\lambda_\odot - \lambda_\Sigma),$$

and the values of the period measured on consecutive days can differ by $6 \cdot 10^{-6}$ sec:

$$P_2 - P_1 = P_0 \frac{d}{dt}(\Delta A_3) \frac{V_{\text{ang}}}{57.3} = P_0 \cdot 10^{-3} \cos \beta_\Sigma \cos \cdot (\lambda_\odot - \lambda_\Sigma).$$

4. Correction for Motion of the Sun about the Solar System Barycenter. The value of this correction can be obtained with separate consideration of the motion of the sun about the COGs of the sun—Jupiter and sun—Saturn systems on the assumption that the sun moves uniformly in a circular orbit in the ecliptic plane. From Fig. 7 we find expressions for the projection of the sun's velocity onto the sun—pulsar line for motion of the sun about the COGs of the systems.

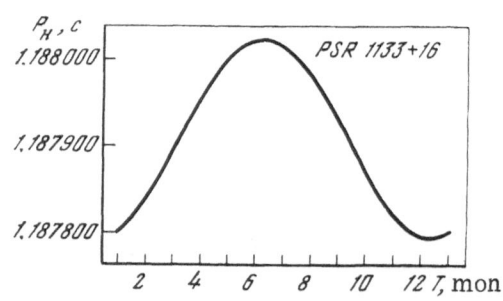

Fig. 6. Annual variation of pulsar period due to the Doppler effect.

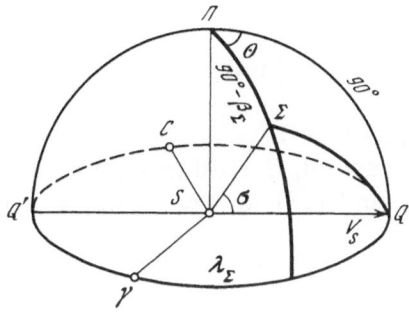

Fig. 7. Motion of the sun about the COG of the sun−Jupiter system.

We consider the spherical triangle $\Pi\Sigma Q$, where the arcs $\Pi\Sigma = 90° - \beta\Sigma$ and $\Pi Q = 90°$, the angle $\theta = l_J + 90° + 180° - \lambda_\Sigma$, and $\cos\sigma = \cos\beta_\Sigma \times \sin(l_J - \lambda_\Sigma)$. The projection of the sun's velocity is

$$V\cos\sigma = V_0\cos\beta_\Sigma \cdot \sin(l_J - \lambda_\Sigma).$$

Similarly, for the sun−Saturn system we have

$$V\cos\sigma = V_0\cos\beta_\Sigma \cdot \sin(l_S - \lambda_\Sigma).$$

The linear velocity V_0 of the sun is equal to 0.0125 km/sec in the case of the sun−Jupiter system and to 0.0027 km/sec in the case of the sun−Saturn system. The required corrections are introduced according to the equations

$$\Delta A_4 = 0.4167 \cdot 10^{-7} \cos\beta_\Sigma \cdot \sin(l_J - \lambda_\Sigma),$$

$$\Delta A_5 = 0.9000 \cdot 10^{-8} \cos\beta_\Sigma \cdot \sin(l_S - \lambda_\Sigma);$$

here l_J and l_S are the heliocentric longitudes of Jupiter and Saturn.

We have thus derived corrections for the influence of all motions of the observer on the period of a pulsar. To obtain the pulsar period in a moving coordinate system it is necessary to correct the barycentric value of the period according to the expression

$$P_0 = P_0\left(1 - \sum_i \Delta A_i\right).$$

The inverse transformation from the observed value of the period P_0 to the barycentric value is realized by correction according to the expression

$$P_0 = P_0\left(1 + \sum_i \Delta A_i\right).$$

We now summarize the expressions for all the derived corrections:

$$\Delta A_1 = -0.1551 \cdot 10^{-5} \cos\varphi \cdot \cos\delta_\Sigma \cdot \sin t,$$

$$\Delta A_2 = 0.4133 \cdot 10^{-7} \cos\beta_\Sigma \cdot \sin(l - \lambda_\Sigma),$$

$$\Delta A_3 = 499.012\, \frac{2\pi}{T}\, \frac{[1 + e\cos(\lambda_\odot - v)]}{\sqrt{1 - e^2}}\, \frac{1}{\cos i}\, \cos\beta_\Sigma \cdot \sin(\lambda_\odot - \lambda_\Sigma - i),$$

$$\Delta A_4 = 0.4167 \cdot 10^{-7} \cos\beta_\Sigma \cdot \sin(l_J - \lambda_\Sigma),$$

$$\Delta A_5 = 0.9000 \cdot 10^{-8} \cos\beta_\Sigma \cdot \sin(l_S - \lambda_\Sigma).$$

LITERATURE CITED

1.   Reference Book of Celestial Mechanics and Astrodynamics [in Russian], Nauka, Moscow
      (1971).
2.   G. M. Clemence and V. Szebenely, Astron. J., 72:1324 (1967).
3.   S. N. Blazhko, Course in Spherical Astronomy [in Russian], Gostekhizdat, Moscow, (1954).
4.   L. Ya. Martynov, Course in Practical Astrophysics [in Russian], Fizmatgiz, Moscow
      (1960).
5.   Astronomical Yearbook of the USSR [in Russian], Nauka, Moscow (1975).

SONIC DIRECTION FINDING IN THE PRESENCE OF STRONG ACOUSTIC INTERFERENCE

V. I. Veksler and E. L. Feinberg

The statistical approach is used to treat the problem of signal detection in the presence of strong noise as the problem of measurement of a small correlation coefficient. A circuit is proposed, based on the coincident operation of several parallel channels with a definite condition assigned to the separate channels. In particular, special attention is given to the polarity-coincidence circuit, which affords a significant advantage over a circuit employing simple addition of the channel currents. The useful output of the proposed circuit increases as the square of the number of channels, and the background level does not depend on the interfering noise intensity. An experiment is described, corroborating the predictions of the theory.

FROM THE EDITOR

This article was written at the beginning of 1943 and was presented at a scientific seminar of the Lebedev Physics Institute (FIAN) in April of the same year. The solution of the problem set forth in the title was based on the following principle stated in the Introduction: "From the physical point of view the problem is to identify the correlated part (sound in the line of sight of the acoustic baseline relative to its source) of an aggregate process, a large part of which comprises totally uncorrelated events (interfering noise, producing incoherent oscillations in different sound receivers)." This principle,* at that time new, reduced the problem to measurement of a small correlation coefficient and, accordingly, to the replacement of linear by nonlinear systems. It engendered a host of applications. For example, a universal correlometer ["a device for measuring (calculating) the values of the correlation coefficient of stationary random processes"] was designed and built according to the circuit diagram in Fig. 4 of the article.†

It is well known that the correlation approach to acoustical and electronic engineering problems was later proposed and developed independently by many authors and has long been universally recognized, even to the point of triviality. The correlation approach was originally used strictly for two-channel devices and under the assumption that the attendant noise was reasonably stationary and (or) the signal spectra were of a special type. Later, however, circuit configurations appeared which were closer to the one described in the present article.

---

*We recall, for example, that the classical works of Rice on the statistical theory of noise immunity for radar devices did not appear until July of 1944 and January of 1945 [1], and then even in the subsequent post-war period they were not immediately available in our country.

† Experiments with a four-channel correlometer of this type and a certain elaboration of the theory are described by Gershman and Feinberg [2]. An application to architectural-acoustics problems has also been published by Gershman [3].

Outside the Soviet Union, a correlator applicable for spectra of any nature and based on coincidence of the signs of fluctuations (polarity-coincidence correlator; see Fig. 1 and § 2 of the present article) appears to have been first proposed and used to record a signal against background noise in the United States in 1952.* The same principles cropped up in many other papers, which of course were written quite independently of the present study.†

One additional remark is called for in regard to the style in which the present article is written and the terminology used therein. The text of the article was geared to scientists, engineers, and technicians in underwater acoustics, to whom the statistical approach, the correlation coefficient concept, and the utilization of nonlinear systems for its measurement (specifically the scheme proposed in the present article for polarity coincidence in parallel channels) were totally unfamiliar. At that time linear ("interference," in the terminology of the article) methods prevailed (and then with subjective detection of the output "by ear"; see the Introduction and Section 1). The authors therefore endeavored to shelter their intended audience from unfamiliar or what proved at first exposure to be intimidating terms, and so wherever posible they replaced the words "correlation" and "absence of correlation" by "coherence at different points" and "incoherence," etc. For the functions $\Gamma(x)$ and $\gamma(x)$ in expression (3) of the article, which are the correlation (more precisely, autocorrelation) functions of the noise and signal, respectively, the authors did not use this term, which was then little known to the intended reader (but is taken quite for granted today).

LITERATURE CITED IN EDITORIAL PREFACE

1.   S. O. Rice, "Mathematical analysis of random noise," Bell Syst. Tech. J., 23:282 (1944); 24;47 (1945).
2.   S. G. Gershman and E. L. Feinberg, "On the measurement of the correlation coefficient," Akust. Zh., 1:326 (1955).
3.   S. G. Gershman, "The correlation coefficient as a criterion of the acoustic quality of a closed room," Zh. Tekh. Fiz., 21:1492 (1951).
4.   J. J. Farlan, Jr., and R. Hills, Jr., Acoustics Res. Lab., Harvard University, Cambridge, Massachusetts, Tech. Memo. No. 27 (September 15, 1952); No. 28 (Novemeber 15, 1952).
5.   S. S. Wolff, J. B. Thomas, and T. R. Williams, IRE Trans. Inf. Theory, IT-8:2 (1962).
6.   R. Hanbury Brown and R. Q. Twiss, Philos. Mag., 45:663 (1954).

INTRODUCTION

The present article is an account of a preliminary study undertaken by the authors for the purpose of finding ways to solve the problem of sonic direction finding (DF) in the presence of acoustic interference that is stronger than the signal.

The methods currently in use clearly fail to give a satisfactory solution of the problem. The better ones are based on the use of a so-called interference system, in which the currents from a large set of sound receivers arrayed along a single acoustic baseline are simply summed. This type of system (see Section 1 for the details) produces considerable enhancement

---

*A two-channel instrument was described in two technical memoranda [4], which were not published and remained inaccessible, but eventually became known through references to them in later publications (see, e.g., [5], in which a comparison is also made with other correlation-measurement techniques).

† A correlator designed specifically for radio-astronomical applications and based simply on a two-channel configuration rather than polarity coincidence was first described in a celebrated paper by Hanbury Brown and Twiss [6].

of a weak target sound if the acoustic baseline is aligned with the true DF line of sight, by comparison with random noise, which merely creates a certain background.

This method is still not sensitive enough, however, and it has an additional feature capable of eliciting major problems, namely the fact that the resultant background depends strongly on the interfering noise level. In real situations this consideration is clearly one of the factors standing in the way of objective detection and necessitating subjective indication. Of course, subjective indication has the advantage that it enables one to make use of the favorable aspects of tonal-coloring differences between the target and interfering noises.

From the general-physical point of view the problem is to identify the correlated part (sound in the line of sight of the acoustic baseline relative to its source) of an aggregate process, a large part of which comprises totally uncorrelated events (interfering noise, producing incoherent oscillations in different sound receivers).

It will be shown below that inference summation, being the only possible method, is not really even the best method of enhancing sound from a source in the line of sight, relative to the interfering noise, which creates disordered currents. In particular, appreciable gains are offered by a conceptually new method — the method of nonlinear signal transformation by means of relays.

In what seems to be the most workable version of the new method, there are two considerations that promise significant advantages over the interference method. First, the excess of the output effect above the background level when a source of weak sound is intercepted by the true line of sight of DF turns out, given the same number, type, and positions of the sound receivers, to be much greater; it increases as the square of the number of receivers, whereas in the interference method it is proportional to the first power of that number. Second, the background induced by the interfering noise is almost totally independent of its intensity and is easily filtered out by simple electronic techniques.

Certain limitations are imposed on this method by fluctuations in the number of coincidences, which keep the optimum number of relays to around eight or ten. However, by combining the method with simple interference summation it is always possible to gain an eight-decibel increase in the discernible difference between the levels of the interfering noise and the target sound by comparison with the ordinary interference method. This estimate ignores the enormous additional advantages afforded by the fact that the background is independent of the noise intensity. This consideration is nonetheless of fundamental significance in the implementation of direction finding.

Below we present the results of a theoretical analysis (Part A: Theory) and an experimental verification (Part B: Experiment) carried out under comparatively unfavorable laboratory conditions with equipment specially adapted for this purpose. These results confirm the validity of the principles underlying the method. In particular, a level difference of 16 to 20 dB can be discerned quite easily with the use of four sound receivers. In our opinion the results obtained here indicate the desirability of commencing further development of the method at the earliest possible date, in particular the design of a ten-channel instrument for testing of the method under laboratory as well as field conditions.

PART A: THEORY

1. Interference Method

The principal method employed at the present time for the detection and direction finding of weak sounds in the presence of strong noise is based on the use of a large number N of sound receivers (hydrophones), which are connected in parallel and arrayed along a certain linear acoustic baseline. In terms of total excited current this kind of system is far more sensitive than a single sound receiver to the occurrence of a weak additional sound source in the DF line of sight.

Here and elsewhere we describe the interfering noise at a given point (we think of it as being emitted, say, by a source out of the line of sight) by a time function $\Psi(t)$, and the weak target sound by a function $\psi(t)$. Their corresponding intensities are I_0 and I: $I_0 = \overline{\Psi^2(t)}$, $I = \overline{\psi^2(t)}$, where the overbar signifies time averaging over a sufficiently long time interval. The difference between the arrival times of the interfering noise at the i-th and at the first receiver is t_i ($t_1 = 0$); the analogous delay time for the weak target sound is t_i'. In the exact line of sight all $t_i' = 0$.

Thus, the total current at the common output of all N receivers connected in parallel is described by the time function (assuming that no distortions other than linear amplification occur in the receiving system)

$$i_S = \sum_{i=1}^{N} [\Psi(t + t_i) + \psi(t + t_i')].\tag{1}$$

An instrument connected to the input performs averaging over a time interval equal to the observation period. This interval is always large in comparison with the periods between sign changes of the functions Ψ and ψ at the highest frequencies significantly represented in the spectrum ($\nu \sim 3$ or 4 kHz). Consequently, a certain rectification of the current usually takes place, and whatever is measured is the time average of a quadratic function of the current. Thus, a measure of the net effect is the quantity

$$\sqrt{\overline{i_S^2}} = \sqrt{\overline{\left\{\sum_{i=1}^{N} [\Psi(t + t_i) + \psi(t + t_i')]\right\}^2}}.\tag{2}$$

If Ψ and ψ describe effectively random sounds (as in the case of ship noise), the average of the product of two such functions taken at different times $\overline{\Psi(t)\Psi(t')}$ has a nonzero value only when the times t and t' are separated by a time interval $|t - t'|$ small in comparison with the characteristic period of the noise amplitude.* We can therefore write, in general,

$$\overline{\Psi(t)\,\Psi(t')} = \overline{\Psi^2(t)}\,\Gamma\left(\frac{|t-t'|}{\theta}\right),$$
$$\overline{\psi(t)\,\psi(t')} = \overline{\psi^2(t)}\,\gamma\left(\frac{|t-t'|}{\vartheta}\right),\tag{3}$$

where the functions $\Gamma(x)$ and $\gamma(x)$ are such that for $x = 0$ they are equal to unity and for values of x greater than unity they rapidly vanish; θ and ϑ are the characteristic periods (at the highest frequencies significant in the spectrum) of the two sounds.

These equations indicate that acoustic oscillations observed at two times sufficiently far apart are essentially mutually independent. For example, noise arriving at two well-spaced receivers will generate incoherent oscillations in them if the source is not situated in the line of sight relative to the receivers. In other words, the receiver currents are incoherent if the relative delay time $|t_i - t_k|$ is large in comparison with a time interval θ, characteristic of the noise, such that all correlation of the acoustic oscillations is lost in that time.

---

* This period is called the "incoherence time." In the general case it can differ for the interfering noise and the target sound.

Squaring and averaging, we obtain

$$\overline{i_S^2} = \sum_{i=1}^{N} \sum_{k=1}^{N} [\overline{\Psi(t+t_i)\,\Psi(t+t_k)} + \overline{2\Psi(t+t_i)\,\psi(t+t_k')} +$$

$$+ \overline{\psi(t+t_i')\,\psi(t+t_k)}] = \overline{\Psi'^2} \sum_{i=1}^{N} \sum_{k=1}^{N} \Gamma\left(\frac{|t_i-t_k|}{\theta}\right) + \overline{\psi^2} \sum_{i=1}^{N} \sum_{k=1}^{N} \gamma\left(\frac{|t_i'-t_k'|}{\vartheta}\right). \qquad (4)$$

The second sum drops out, because the two sounds are mutually independent.

If the receivers are well spaced, the interference-induced currents in them are incoherent. In the first sum, therefore, only terms with i = k remain, so that $\Gamma = 1$. Assuming that the target sound is exceedingly weak and taking the approximate square root, we obtain

$$\sqrt{\overline{i_S^2}} = \sqrt{N\overline{\Psi'^2}}\left(1 + \frac{1}{2N}\frac{\overline{\psi^2}}{\overline{\Psi'^2}}F\right), \qquad F = \sum_{i=1}^{N} \sum_{k=1}^{N} \gamma\left(\frac{|t_i'-t_k'|}{\vartheta}\right). \qquad (5)$$

Here the function F characterizes the way in which the addition to the background $(N\overline{\Psi^2})^{1/2}$ varies as the position of the weak sound source is varied relative to the baseline (the delay times t_i' are determined exclusively by the relative positions of the sound receivers and source). In the exact line of sight the currents excited in all receivers are coherent, and $\gamma = 1$. Therefore (introducing the intensity in place of the amplitude), for the exact line of sight we have

$$\sqrt{\overline{i_S^2}} = \sqrt{NI_0}\left(1 + \frac{N}{2}\frac{I}{I_0}\right). \qquad (6)$$

(We are assuming that $NI/I_0 \ll 1$.)

Thus, the output current depends on the intensity of the weak target sound. By increasing the number of receivers it is possible to detect weaker and weaker sounds. Requiring, for example, that the target sound induce a current 50% above the background level, we obtain

$$\frac{N}{2}\frac{I}{I_0} = \frac{1}{2}, \qquad \frac{I}{I_0} = N, \qquad (7)$$

i.e., the discernible difference in levels, measured in decibels, is equal to 10 log N under the given requirement, where N is the number of receivers. For N = 10 this condition yields 10 dB, and for N = 100 we obtain 20 dB.

We have quite intentionally analyzed this now-commonplace situation in such detail to provide a consistent basis of comparison between the interference method of sonic direction finding and the relay method described below.

2. RELAY METHOD

Let us consider a circuit configuration in which the currents excited in the sound receivers are not simply summed, as in the interference method, but are transmitted from each receiver to a separate relay, which operates when a definite condition is met. Let us suppose also that the recording instrument at the output gives a reading only when all relays operate simultaneously* (Fig. 1). In this case the statistical calculation of the effect is altogether different. We propose to show that much greater enhancement of a weak sound is attainable by this approach when the sound source is in the DF line of sight and the noise source is not.

_____

* By means of an appended master relay.

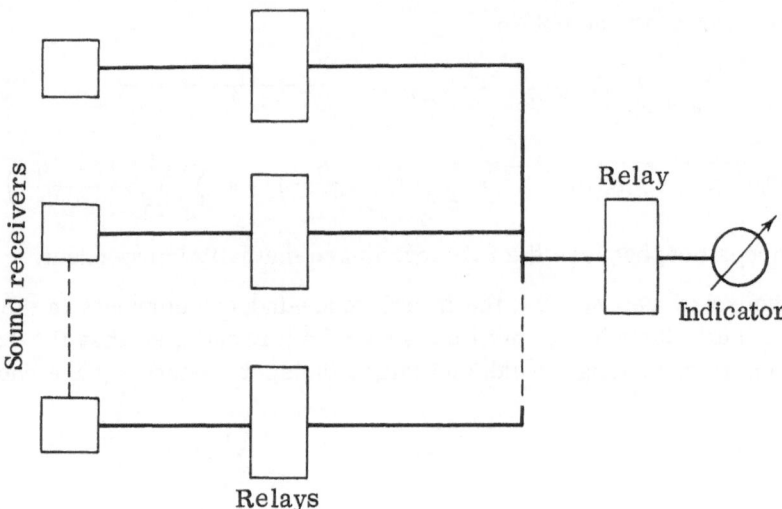

Fig. 1

Let each relay operate only if the acoustic oscillation at the given instant satisfies a definite requirement, say, that the total value of the deviation $\Psi + \psi$ exceed a certain threshold level V_k. This is the simplest possible condition (particularly insofar as we are assuming that $V_k \approx 0$) and is clearly the best suited to our task. As another possible alternative we have also considered a radio-engineering circuit characterized by the trigger condition

$$\frac{d}{dt}(\Psi + \psi) > A[1 + B^2(\Psi + \psi)^2], \tag{8}$$

in which A and B are constants. This condition, which is met in other applications, is less favorable in the present situation. In general, any trigger condition of the following type is possible:

$$a < F\left(\Psi + \psi, \ \frac{d}{dt}(\Psi + \psi), \ \frac{d^2}{dt^2}(\Psi + \psi), \dots\right) < b. \tag{9}$$

Thus, we choose the condition

$$\Psi + \psi > V_k, \tag{10}$$

assuming as a rule that $V_k^2 \ll \overline{(\Psi + \psi)^2}$, i.e., that a relay will operate for almost any positive value of the amplitude.

It is obvious that, since a sound wave does not contain a constant component, the probability of operation of one channel is the same at every instant and is approximately equal to $w_1^0 = \frac{1}{2}$.

If the noise in two channels is completely independent, the probability of simultaneous operation of both channels is $w_2^0 = (w_1^0)^2 = \frac{1}{4}$, and in general the probability of simultaneous operation of N channels is $w_N^0 = (w_1^0)^N = 2^{-N}$.

Suppose that the noise is made up of a weak sound that is, to some extent at least, coherent in different channels (i.e., comes from a source located close to the exact line of sight of DF). At times when the weak sound amplitude is positive in a given channel the operation probability of that channel increases by a certain small amount; when the amplitude is negative, this probability decreases. In general the probability becomes slightly time-dependent and equal to

$$w_1 = w_1^0 + \delta (t). \tag{11}$$

Clearly, $\overline{\delta (t)} = 0$, since for one channel the admixing of sound does not render the noise "less random." In different channels the increment function $\delta (t)$ [which, of course, is associated in some way with the target sound $\psi (t)$] has different values $\delta (t + t_i')$ in general. Only in the exact line of sight will δ be identical in all channels.

Now the simultaneous-operation probability for N channels* is

$$w_N = (w_1^0 + \delta (t)) (w_1^0 + \delta (t + t_2')) \ldots (w_1^0 + \delta (t + t_N')) = (w_1^0)^N \prod_{i=1}^N \left(1 + \frac{\delta (t + t_i')}{w_1^0}\right). \tag{12}$$

The output current or potential will be proportional to this N-channel simultaneous-operation probability. The average result is obtained by averaging w_N over the observation period. Inasmuch as $\delta \ll w_1^0$, the result can be limited to the lowest powers of the ratio δ / w_1^0 after cross-multiplication:

$$\bar{i}_R = \overline{w}_N = (w_1^0)^N \left(1 + \sum_{i=1}^N \overline{\frac{\delta (t + t_i')}{w_1^0}} + \sum_{i \neq k=1}^N \overline{\frac{\delta (t + t_i') \delta (t + t_k')}{(w_1^0)^2}} + \ldots \right). \tag{13}$$

In the exact line of sight, also making use of the fact that $\overline{\delta (t + t_i')} = 0$, we have

$$\bar{i}_{R, \max} \sim (w_1^0)^N \left(1 + \frac{N (N - 1)}{2} \frac{\overline{\delta^2}}{(w_1^0)^2}\right). \tag{14}$$

Thus, the resulting excess above the background level in the DF line of sight, by contrast with the interference method, increases as $N (N - 1)$ rather than as N, i.e., for large N it will be significantly greater. This result forms the basis of the relay method. All that is required for a particular circuit configuration is to determine $(w_1^0)^N$ (which in our problem is roughly equal to 2^{-N} and so does not depend on the interference level) and to express $\overline{\delta^2} / (w_1^0)^2$ in terms of the intensities I and I_0. We now explore the overall problem in greater detail.

In the case of completely random noise, by representing the acoustic process as random events we can consider the quantities Ψ and ψ to be distributed according to laws of chance. Consequently, the probability that Ψ will assume values between Ψ and $\Psi + d\Psi$ at a given time is equal to

$$P (\Psi) d\Psi = A \exp (-\Psi^2/2\overline{\Psi^2}) d\Psi, \quad A = (2\pi\overline{\Psi^2})^{-1/2}, \quad I_0 = \overline{\Psi^2}. \tag{15}$$

Accordingly,

$$p (\psi) d\psi = a \exp (-\psi^2/2 \overline{\psi^2}) d\psi, \quad a = (2\pi\overline{\psi^2})^{-1/2}, \quad I = \overline{\psi^2}. \tag{16}$$

Operation of a particular channel is realized every time that the sum $\Psi + \psi$ exceeds the specified threshold V_k, i.e., when $\Psi > V_k - \psi$. The probability of operation of a single channel at a given time t when ψ has a certain definite value, equal to $\psi (t + t_i')$ in the i-th channel, is equal to the integral

$$\int_{V_k - \psi (t + t_i')}^{\infty} P (\Psi) d\Psi. \tag{17}$$

---

*Namely the probability that $\Psi > V_k - \psi (t + t_1')$ in the first channel, $\Psi > V_k - \psi (t + t_2')$ in the second, etc. (as the probability of coincidence for independent events).

The probability that all N channels will operate simultaneously at a given time (if the noise Ψ in different channels is independent) is equal to the product of the individual probabilities:

$$\prod_{i=1}^{N} \int_{V_k - \psi(t+t_i')}^{\infty} P(\Psi)\, d\Psi. \tag{18}$$

The average current or average potential at the output of the instrument is obtained by averaging the latter expression over all values assumed by the function ψ during the observation period. Instead, averaging can be realized in the usual way by means of the function $p(\psi)$. If the source in this case is situated close to the DF line of sight, then the functions $\psi(t + t_i')$ are not independent in different channels, rather they are related in a very definite way. Let us denote, for example, $\psi(t + t_i')$ in the first channel by $\psi = \psi(t)$. Then in the other channels $\psi(t + t_i')$ is expressed in terms of $\psi(t)$ and a delay time. Only averaging with respect to the values of $\psi(t)$ remains. The average probability of simultaneous operation is

$$\bar{w}_N = \int_{-\infty}^{+\infty} p(\psi)\, d\psi \prod_{i=1}^{N} \int_{V_k - \psi(t+t_i')}^{\infty} P(\Psi)\, d\Psi. \tag{19}$$

In particular, in the exact line of sight, where $\psi(t + t_i') = \psi(t)$,

$$\bar{w}_N = \int_{-\infty}^{\infty} p(\psi)\, d\psi \left[\int_{V_k - \psi(t)}^{\infty} P(\Psi)\, d\Psi \right]^N. \tag{20}$$

We now discuss what is for us the most important case, namely an exceedingly weak target sound ($I \ll I_0$); we also assume that $V_k \ll \sqrt{I_0}$. In this case

$$\left[\int_{V_k - \psi}^{\infty} P(\Psi)\, d\Psi \right]^N \approx \left[\frac{1}{2} - A(V_k - \psi) \right]^N =$$
$$= \frac{1}{2^N} \left\{ 1 - 2NA(V_k - \psi) + \frac{N(N-1)}{2} [2A(V_k - \psi)]^2 + \ldots \right\}. \tag{21}$$

Substituting this expression into (20) and integrating with respect to ψ, we obtain

$$\bar{w}_N \approx \left(\frac{1}{2} \right)^N \left(1 - \frac{2NV_k}{\sqrt{2\pi I_0}} \right) \left[1 + \frac{N(N-1)}{\pi(1 - 2NV_k/\sqrt{2\pi I_0})} \frac{I}{I_0} \right]. \tag{22}$$

Thus, in actuality w_N and, hence, the potential (or current) at the output has the form given by (14). The excess above the background level due to weak sound in the line of sight depends quadratically on the number of receivers. It is easily shown that near the exact line of sight, in which case it is necessary to discern different functions $\psi(t + t_i')$ in (19), the successive expansion of the Gaussian integral with respect to the small lower limit yields a factor F exactly the same as in (5) in the additional term containing I/I_0. Consequently, the DF curve retains the same form as in the interference method, but the absolute dimensions change.

A second important feature of this version of the relay method is the fact that for $V_k = 0$ the background is totally independent of the intensity of the interfering noise. In practice, however, V_k is always slightly different from zero. If V_k is positive, then the background increases with the noise intensity I_0, while for negative V_k the background diminishes with noise amplification.

If we put $V_k = 0$, then in the DF line of sight

$$\bar{w}_N = \frac{1}{2^N} \left[1 + \frac{N(N-1)}{\pi} \frac{I}{I_0} \right]. \tag{23}$$

Comparing this expression with the expression (6) for the interference method, we see that with the same requirement for 50% excess above the background level the intensity ratio is

$$\frac{I_0}{I} = \frac{2}{\pi} N (N - 1). \tag{24}$$

Accordingly, the discernible difference between levels in decibels is $10 \log (I_0/I) = 10 \log N + 10 \log [2(N-1)/\pi]$. For N = 10 this condition yields roughly an additional 8 dB over the interference method (18 dB instead of 10 dB). Additional possibilities are afforded by the negligible dependence of the background on I_0, making it possible to easily record a relative excess above the background level much smaller than 50% (see Part B: Experiment).

It will be shown below that a limitation is imposed on the admissible number N of relay channels by the impossibility of making the observation period too long, so that the role of fluctuations becomes significant for large N.

3. Role of Fluctuations at the Output

It is evident from expression (23) that the relay method leads to considerable attenuation of the output current (by the factor 2^{-N}, which is $\sim 10^{-3}$ for N = 10). Attenuation of the current or potential per se should present no great danger, because any amplifying device could be used. It is significant, however, that separate pulses arrive in the terminal part as a result of each simultaneous operation of the entire relay system. An increase in the number of relay channels makes such simultaneous operations extremely rare events. For large N, therefore, the number of pulses arriving in the observation period will be so small that fluctuations in their number can clearly become exceedingly significant.

Pulsewidth at the Output. The duration of the time intervals spanned by simultaneous operation of all relays assumes very diverse values, first, because the operating times of each channel fluctuate and, second, because these time intervals for different channels overlap in all different ways. At any rate, obviously, for large N the times τ_N of simultaneous operation are much smaller than the operating times τ_1 in each separate channel.

Suppose that the distribution of the operating times for one relay is described by the function $n(\tau_1)$, so that the total duration of all operating time intervals for a single channel per unit time is

$$\int_0^\infty \tau_1 n (\tau_1) \, d\tau_1. \tag{25}$$

But for $V_k = 0$, as we know, the foregoing quantity is smaller by one-half. We therefore obtain the normalization condition for $n(\tau_1)$:

$$\int_0^\infty \tau_1 n (\tau_1) \, d\tau_1 = {}^1\!/_2. \tag{26}$$

If sound of frequency $\nu = 1/T$ were delivered to the input of the instrument, we would have $\tau_1 = {}^1\!/_2 T$, $n(\tau_1) = \nu\delta(\tau_1 - {}^1\!/_2 T)$, where $\delta(x)$ is a delta function, equal to zero whenever x has a nonzero value, $\int_{-\infty}^\infty \delta(x) \, dx = 1$. We can make the approximate assumption that a certain effective frequency ν_{ef} exists, characterizing the number of sign changes of Ψ or ψ per unit time (more precisely, ν_{ef} is equal to half this number). It can be shown that ν_{ef} is of the order of the highest frequency significantly represented in the sound spectrum. This replacement of the function $n(\tau_1)$ would imply that we are neglecting fluctuations of the values of τ_1. For the time being we shall not make this simplifying assumption.

We determine the probability that a given channel will operate at time t and at all times between t and $t + \tau_N$ very close to t. We first consider only intervals of a given duration τ_1. The required probability is equal to the probability of being in one of the intervals of duration τ_1 [i.e., probability $\tau_1 n(\tau_1) d\tau_1$] multiplied by the probability of being at a point situated at a distance greater than τ_N from the end of that interval (probability $1 - \tau_N \tau_1^{-1}$). Consequently, for pulses of width τ_1 the probability that operation in one channel will take place and last at least from t to $t + \tau_N$ (and possibly even longer) is

$$\tau_1 n(\tau_1)\, d\tau_1 \left(1 - \tau_N/\tau_1\right). \tag{27}$$

The total probability (irrespective of the widths of the pulses themselves) of such a resultant operation lasting at least τ_N is

$$\int_{\tau_N}^{\infty} \tau_1 n(\tau_1)\, d\tau_1 \left(1 - \frac{\tau_N}{\tau_1}\right). \tag{28}$$

The probability that simultaneous operation in all channels will last from t to $t + \tau_N$ (and possibly longer) is

$$\left[\int_{\tau_N}^{\infty} \tau_1 n(\tau_1)\, d\tau_1 \left(1 - \frac{\tau_N}{\tau_1}\right)\right]^N. \tag{29}$$

The probability that operation of all N relays will last exactly a time τ_N between τ_N and $\tau_N + d\tau_N$ (but not longer) is equal to minus the differential of this expression with respect to τ_N:

$$w_N^{(\tau_N)}\, d\tau_N = N \left[\int_{\tau_N}^{\infty} n(\tau_1)(\tau_1 - \tau_N)\, d\tau_1\right]^{N-1} \int_{\tau_N}^{\infty} n(\tau_1)\, d\tau_1\, d\tau_N. \tag{30}$$

This equation gives the distribution function of the coincidence intervals at the output when the distribution of coincidence intervals in one channel is known.

If N is large and, as should be expected and is indeed the case, $\tau_N \ll \bar{\tau}_1 = \tau_0$, then the lower limit of the integral can be set equal to zero, i.e., for example, the expansion of the Gaussian integral with respect to a small lower limit can be restricted to the first term if $n(\tau_1)$ is a Gaussian error function. Then, using (26), we obtain

$$w_N^{(\tau_N)}\, d\tau_N = N n_1 \left(\frac{1}{2} - n_1 \tau_N\right)^{N-1} d\tau_N. \tag{31}$$

Here n_1 is the total average number of single-channel operations, which is equal to $\nu_{ef} = (2\bar{\tau}_1)^{-1}$. Consequently,

$$w_N^{(\tau_N)}\, d\tau_N = \frac{N}{2^N} \left(1 - \frac{\tau_N}{\bar{\tau}_1}\right)^{N-1} \frac{d\tau_N}{\bar{\tau}_1} \approx \frac{1}{2^N} \frac{N}{\bar{\tau}_1} \left[1 - (N-1)\frac{\tau_N}{\bar{\tau}_1}\right] d\tau_N. \tag{32}$$

Here the first factor 2^{-N} characterizes the probability that coincidence will take place in general, while the rest indicate the probability that coincidence will last a time τ_N. Therefore, the distribution of coincidence intervals is given by the functions

$$w_N(\tau_N)\,d\tau_N \approx \frac{N}{\tau_1}\left[1 - (N-1)\frac{\tau_N}{\tau_1}\right]d\tau_N \approx \frac{N}{\tau_1}\exp\left(-N\frac{\tau_N}{\tau_1}\right)d\tau_N. \tag{33}$$

Thus, the distribution of coincidence intervals is extremely flat. The average coincidence interval is

$$\bar{\tau}_N = \bar{\tau}_1/N. \tag{34}$$

For a small number N of channels it is necessary to inject definite assumptions in the form of the function $n(\tau_1)$. We assume that all coincidence intervals in one channel are identical and equal to τ_0, i.e., that $n(\tau_1)$ is a delta-type function. From (30) we then readily deduce an interval distribution function in the form of the same function (32), with average value

$$\bar{\tau}_N = \bar{\tau}_1/(N+1) = \tau_0/(N+1), \quad \tau_0 = {}^1\!/_2\nu_{\mathrm{ef}}. \tag{35}$$

This is the value that we shall use.

It can be shown that the spectrum of a pulse of width τ_N contains all frequencies in equal measure up to a frequency of order $1/\tau_N$, along with higher frequencies in much smaller measure. Consequently, the spectrum in the output part of the circuit extends from low frequencies to frequencies of order $N\nu_{\mathrm{ef}}$, where ν_{ef} is the effective (highest) frequency of the primary spectrum. This condition determines the requirements on the frequency response of the circuit in its terminal section.

Number of Pulses at the Output. We assume that at the first instant of simultaneous operation of all relays a certain constant potential V_0 is established in the terminal section of the circuit, vanishing at the end of operation. The number of pulses at the output in this case is easily determined on the basis of the following considerations.

The probability of simultaneous operation at a given instant is 2^{-N}. Consequently, the average output potential is $2^{-N}V_0$. But, on the other hand, it is made up of a certain number n of pulses per unit time, each of average width $\tau N = \tau_0/(N+1)$. Therefore,

$$n_N \bar{\tau}_N = 2^{-N}, \quad n_N = \frac{N+1}{2^N}\frac{1}{\tau_0} = \frac{N+1}{2^{N-1}}\nu_{\mathrm{ef}}. \tag{36}$$

Thus, the number of pulses per unit time diminishes rapidly as the number of channels is increased. For N = 10 and ν_{ef} = 4 kHz we have n_N = 80. The fluctuations of this number also grow proportionately larger.

Voltage Fluctuations at the Output. We now deduce a complete expression for the output potential fluctuations with regard for fluctuations of the duration and number of coincidences as well as the finiteness of the observation time. Here, as before, we assume that the output potential is equal to a constant value V_0 throughout the entire time of simultaneous operation of all relays and is equal to zero the rest of the time.

The calculation of the circuit operation probability in Section 2 may be regarded as a calculation of the average value of a product of functions, which we denote by the symbol $\delta^{a,\,\Psi+\psi,\,b}$. Each of these functions is equal to unity for $a < \Psi + \psi < b$ and is equal to zero for other values of $\Psi + \psi$. In our situation $a = 0$ and $b = \infty$. Thus, formal averaging by means of the functions $p(\psi)$ and $P(\Psi)$ [see (15) and (16)] yields the average output potential

$$\bar{V}_N = V_0 \int_{-\infty}^{+\infty} P(\Psi_1)P(\Psi_2)\ldots P(\Psi_N)\,d\Psi_1\,d\Psi_2\ldots d\Psi_N \int_{-\infty}^{+\infty} p(\psi)\,d\psi\,\delta^{0,\,\Psi_1+\psi,\,\infty}\delta^{0,\,\Psi_2+\psi,\,\infty}\ldots\delta^{0,\,\Psi_N+\psi,\,\infty}. \tag{37}$$

The rms fluctuation of V_N is

$$\Delta V_N = \sqrt{\overline{(V_N - \overline{V}_N)^2}} = \sqrt{\overline{V_N^2} - \overline{V}_N^2}. \tag{38}$$

Here it is essential to take into consideration that

$$\{\delta^{0,\, x,\, \infty}\}^2 \equiv \delta^{0,\, x,\, \infty}. \tag{39}$$

Since we are calculating mainly the background fluctuations, we can put $\psi = 0$ in the symbols $\delta^{0,\, \Psi + \psi,\, \infty}$. We then readily infer with the aid of (39) that

$$\overline{V_N^2} = V_0 \overline{V}_N. \tag{40}$$

and, since $\overline{V}_N = 2^{-N} V_0$, we have $\sqrt{\overline{V_N^2} - \overline{V}_N^2} = \dfrac{V_0}{2^{N/2}} \sqrt{1 - 2^{-N}}$. From this result we deduce for the relative fluctuation

$$\sqrt{\overline{V_N^2} - \overline{V}_N^2} \Big/ \overline{V}_N = 2^{N/2} \sqrt{1 - 2^{-N}}. \tag{41}$$

The spectrum of these fluctuations is determined by the spectrum of the individual pulses, i.e., it is very high-frequency (see the preceding section). Consequently, if the indicator connected to the output has a sufficiently large time constant rC, only a small part of this spectrum will be represented in the indicator readings. For example, if we assume that all pulses have the same width τ_N and that the average number of them is $n_N = [(N + 1)/2^{N-1}] \nu_{ef}$, then in the time rC the average number of pulse arrivals is $\overline{n} = rC\nu_{ef}(N + 1)/2^{N-1}$, and the rms deviation of these pulses is equal to the square root of the indicated number, because we know that $\overline{\Delta n} = (\overline{n^2} - \overline{n}^2)^{1/2} = \sqrt{\overline{n}}$. The relative fluctuation is therefore

$$\frac{\overline{\Delta n}}{\overline{n}} = \frac{2^{(N-1)/2}}{\sqrt{rC\nu_{ef}(N + 1)}}. \tag{42}$$

If we consider a certain specific indicator hookup arrangement, for example, as shown in Fig. 2, where T is an amplifier tube at the output, then on the basis of (41) we can calculate the fluctuation of the milliammeter readings in the general form $(\overline{i^2} - \overline{i}^2)^{1/2}$, where i is the current through the meter. In order of magnitude we obtain the same expression (42) [the factor $(1 - 2^{-N})^{1/2} \approx 1$ must be dropped everywhere]. Using the expression for ΔV_N, we finally arrive at the following equation for the relative fluctuations of the observed current in the indicator:

$$\delta = \sqrt{\frac{\overline{V^2} - \overline{V}^2}{\overline{V}^2}} \sqrt{\frac{\tau_N}{rC}} = \frac{2^{(N-1)/2}}{\sqrt{rC\nu_{ef}(N + 1)}}. \tag{43}$$

4.   Deciding the Number of Relays

The results of Sections 2 and 3 enable us to solve the problem of the admissible number of channels containing independent relays with allowance for fluctuations.

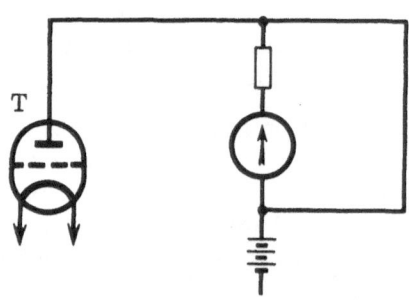

Fig. 2

It is clear that the relative increase observed in the output potential in the line of sight must be greater than the fluctuations of that potential, i.e., the following condition must be satisfied:

$$\frac{N(N-1)}{\pi}\frac{I}{I_0} > \frac{2^{(N-1)/2}}{\sqrt{rC\nu_{ef}(N+1)}}. \tag{44}$$

Consequently, for the observation of weak sounds it is necessary to use a lagging indicator and to work with the high-frequency components of the incoming signal. We take $rC = 1$ sec and $\nu_{ef} = 4$ kHz (it will become apparent that the dependence on the values of these quantities is not very critical). In this case the discernible level difference in decibels is

$$10 \log \frac{I_0}{I} < 10 \log\left[\frac{60}{\pi}\frac{N(N-1)\sqrt{N+1}}{2^{(N-1)/2}}\right]. \tag{45}$$

This result yields approximately the following values for the minimum discernible level difference:

N	2	4	6	8	10
$10 \log (I_0/I)$, dB	17	22	26	24	24

Thus, it is impractical to use a very large number of channels. The optimum appears to be $N \sim 10$.

5. Combined Method

In view of the fact that fluctuations preclude the use of too many independent relays, a practical way to further increase the DF sensitivity is to incorporate the advantages afforded by the conventional interference method.

Let there be given a total of Nm sound receivers connected in parallel groups of m each, with each group connected in series with one of N independent relays (Fig. 3). This arrangement combines simple summation with the relay method. We consider only the exact line of sight of DF.

There arrives in each of the N relay channels, first, random interfering noise, which is summed over m receivers, and in each of the latter it is described by the function

$$P(\Psi)d\Psi = \frac{1}{\sqrt{2\pi I_0}}\exp\left(-\frac{\Psi^2}{2I_0}\right)d\Psi, \qquad I_0 = \overline{\Psi^2}. \tag{46}$$

But the sum of m random variables is itself a random variable, the rms value of which is

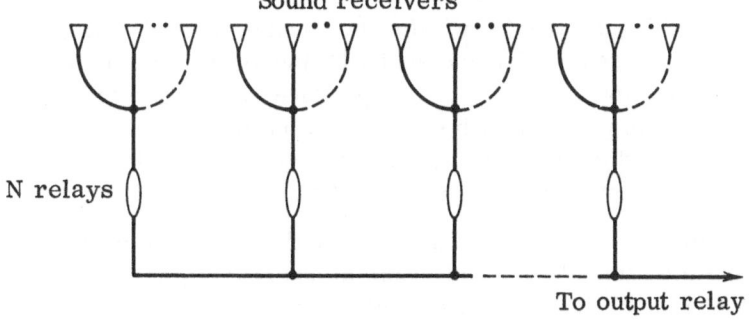

Sound receivers

N relays

To output relay

Fig. 3

larger by a factor $\sqrt{m}$. Consequently, in each of the N relays the noise is described by the function

$$P_m(\Psi)\,d\Psi = \frac{1}{\sqrt{2\pi I_0'}}\exp\left(-\frac{\Psi^2}{2I_0'}\right)d\Psi, \qquad I_0' = mI_0. \tag{47}$$

Second, each relay receives the target sound. In the exact line of sight it is identical in all m receivers connected to the relay. At the input of each relay, therefore, we have

$$p_m(\psi)\,d\psi = \frac{1}{\sqrt{2\pi I'}}\exp\left(-\frac{\psi^2}{2I'}\right), \qquad I' = m^2 I. \tag{48}$$

Thus, we can analyze the previous circuit of N relays with appropriately modified values of the intensities. In place of (23), in the DF line of sight we have the following at the output of such a combined circuit:

$$\bar{w}_{Nm} = \left(\frac{1}{2}\right)^N\left[1 + \frac{N(N-1)m}{\pi}\frac{I}{I_0}\right]. \tag{49}$$

If, on the other hand, instead of combining with the relay method we apply simple interference summation to all Nm receivers, then, replacing N by Nm in (6), we obtain

$$\sqrt{\overline{i_s^2}} = \sqrt{NmI_0}\left(1 + \frac{Nm}{2}\frac{I}{I_0}\right). \tag{50}$$

Thus, the excess above the background level can always be increased by a factor $(2/\pi)\times(N-1)$ relative to that obtained by the conventional interference method if the existing number of receivers is partitioned into N groups of parallel-connected receivers and each group is connected in series with one of the independent relays of a coincidence circuit. Specifying the same excess above the background level, we thus obtain a gain of $10\log[2(N-1)/\pi]$ dB. For N = 10 this result gives an 8-dB increase in sensitivity even without reference to the advantages afforded by nondependence of the background on the noise intensity.

6. Additional Remarks

The above-analyzed version of the relay method, of course, by no means exhausts all the possibilities. We note first of all that the entire discussion has been concerned with electronic relays and, accordingly, the output potential and current. This choise is dictated, on the one hand, by the fact that at the frequencies involved such relays are clearly the most suitable. On the other hand, the choice was necessary for greater definiteness. However, the statistical characteristics of a relay circuit are a general property and would be just as available with the use of mechanical relays. Only the terminology would have to be modified somewhat.

A variety of possibilities exists within the framework of the radio-engineering approach. For example, the sensitivity can easily be increased $\sqrt{2}$-fold by a minor elaboration. Thus, under the condition $\Psi + \psi > V_k$ only positive amplitudes are utilized. If an additional relay is connected in parallel with each relay, but operates under a different condition, $\Psi + \psi < V_k$, then, as we readily perceive, the excess above the background level is twice as great, while the fluctuations are only $\sqrt{2}$ times greater.

On the other hand, V_k could be made not too small, for example, by taking $V_k = -\tfrac{1}{2}\sqrt{I_0}$. Then the excess above the background level would be decreased, but not very appreciable (1/2 to 1/3), whereas, most important, the background would be very greatly increased [by $(3/2)^N$], the relative fluctuations would be diminished accordingly, and so the opportunity would be presented for a major increase in the number of relays. In this case, however, the background would become greatly dependent on the interfering noise intensity, making this whole version unsuitable from the practical standpoint. It is evident in this example, however, that the limitation on the number of channels due to fluctuations need not be viewed as an inherent attribute of the relay method.

PART B: EXPERIMENT

To obtain approximate corroboration of the principles of the relay method we have conducted preliminary experiments with the use of apparatus having a different function, but adapted according to the circuit of Fig. 4. The general configuration of the apparatus is shown schematically in Fig. 5. (We were assisted in the experiments by P. A. Cherenkov.)

A phonograph recording of underwater noise from a source of practical interest was played back through an amplifier and powerful dynamic speaker.

The noise level was varied from 65 to 102 dB by varying the gain. This noise was received by four randomly spaced piezoelectric microphones of the Brach type, playing the role of interference and forming the background against which the weak target sound was to be detected. The four micrphones were connected to four respective channels of a relay device (see Fig. 4) implemented according to a coincidence circuit. The output indicator was a microammeter, from which the output readings were taken.

A large negative bias (~35 V) was applied to the grid of the amplifier tube in the terminal section of the instrument. This measure cut out a large portion of the background. Thus, the reading of the output microammeter did not indicate the full value of the output current (which would be proportional to $\bar{V}_N$; see Part A: Theory), rather only a small fraction of it.

The input of one of the channels was connected to the input of a noise meter with its built-in microphone removed. Thus, the noise-meter readings gave a direct representation of the noise level sensed by the microphone of that channel. Since the microphone normally connected to the noise meter and our piezoelectric microphones can differ somewhat in sensitivity, the absolute noise level in decibels read on the noise-meter scale could contain a certain error, but the level difference in decibels determined in this way may be judged sufficiently reliable.

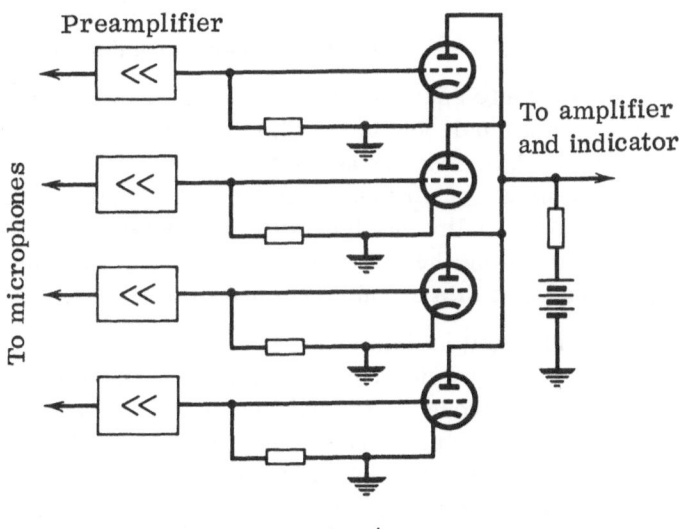

Fig. 4

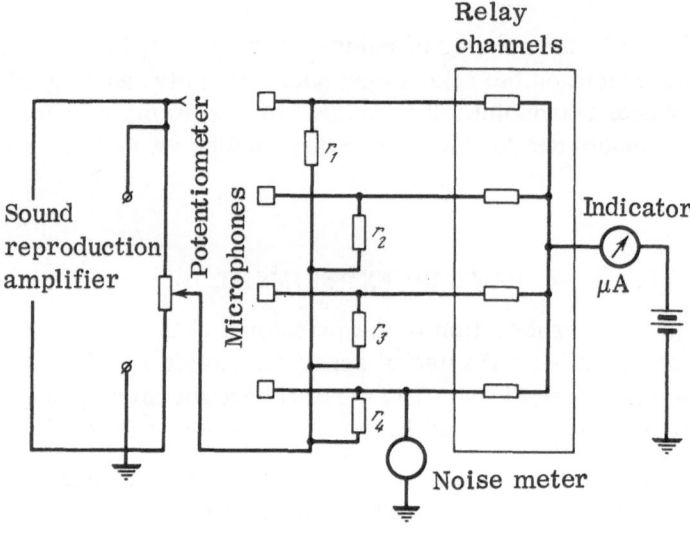

Fig. 5.

We first ran a special test for complete incoherence of the noise-induced currents in all four microphones. For this purpose we eliminated two channels, leaving first one and then the other pair in place, and then moved one of the microphones of the active pair, whereupon we were able to confirm the invariance of the output microammeter readings for all relative displacements of the microphones of the given pair. This means that the microammeter readings correspond to random background and that the noise source is not situated in the DF line of sight relative to either pair of microphones.

As stated above (see Part A: Theory), if the critical operating potential V_k of one relay were exactly equal to zero, the background would be totally independent of the noise level. Measurements showed that the dependence of the background on the noise level is in fact very weak.

Certain difficulties arose in connection with the problem of generating a weak target sound that would arrive simultaneously in all four channels. On the one hand, the acoustics of the room in which the experiments were conducted precluded the possibility of creating a plane sound wave. On the other hand, the microphones at our disposal had several resonance frequencies at 1, 2 kHz, etc. Consequently, small disparities in them produced large discrepancies in the phase shifts. For these reasons we supplied the "target sound" electrically, as follows.

As is shown in Fig. 5, the dynamic speaker was connected through a potentiometer. Electrical oscillations were taken from a small section of this potentiometer, thus reproducing the noise of the same source, and were delivered through large intermediate resistances r_1-r_4 ($\sim 5 \cdot 10^6 \ \Omega$) simultaneously to the inputs of all four channels. This setup simulated sound emitted by a source situated in the exact line of sight with zero delay in different channels relative to one another. By disconnecting the dynamic speaker from the potentiometer it was possible to determine the noise level of this simulated sound directly from the noise meter.

The measurements were performed as follows. First, connecting only the speaker (i.e., setting the potentiometer slide at zero), we determined the interfering noise level in decibels from the noise meter and read the background noise induced at the instrument output from the microammeter at the output of the coincidence circuit. We then moved the potentiometer slide to set a particular level of "weak target sound." At this time the noise meter indicated practically no change, whereas the output microammeter indicated a new current value, which therefore corresponded to the current obtained in the presence of target sound emitted by a source in the DF line of sight. The speaker was then disconnected, and the "target sound" level determined from the noise level.*

Because of the fast response of the noise meter, inadequate uniformity on the part of the sound playback produced large oscillations of the noise-meter pointer. The readings were therefore made within ±1-2 dB error limits. The microammeter readings fluctuated accordingly.

Even with the instrument used for the measurements it turned out that sound with a level 15 to 20 dB (or more) above the interference level was detectable.

The instrument used in our work was not designed for this purpose and certainly did not optimally meet the stated problem. In particular, its main drawbacks were low transmissivity for high frequencies and excessive transmissivity in the low-frequency range.

ADDENDUM

In many sonic direction-finding systems currently in use, at first glance, a different basis is employed for noise abatement. Specifically, a large number of receivers are spaced as close together as possible, making it impossible to speak in terms of independent action of the interfering noise in different receivers. It is assumed in this connection that if it were possible to fill up a certain area solidly with microphones or, even better, to build one enormous microphone with a diaphragm of large diameter in comparison with the wavelength, the action of the noise at different points of this diaphragm would be mutually compensating, and only the pure useful signal would remain. It is easily shown, however, that the end result turns out to be the same here as in the case analyzed above.

We now consider the following ideal limiting case: one square diaphragm of area $S = L^2$, where $L \gg l$, l is the "coherence length," which is related to the coherence time θ by the equation $l = c\theta$ (c is the sound velocity) and is of order $1/k$, $k = 2\pi/\lambda$, and λ is the effective wavelength.

We denote by $\varphi(t)$ the instantaneous value of the pressure created at a given point of the diaphragm by a plane wave from a weak target source. If the source is situated exactly in the line of sight, then φ is identical for all points of the diaphragm. Conversely, the pressure due to interfering noise differs for different points of the diaphragm and depends on the rectangular coordinates x, y in its plane: $\Psi = \Psi(x, y)$. The total pressure experienced by the entire diaphragm at time t is

$$\Pi(t) = \int_0^L dx \int_0^L dy \, [\Psi(t; x, y) + \varphi(t)].$$

The effective value (after rectification) at the output of the entire instrument is proportional to $(\overline{[\Pi(t)]^2})^{1/2}$, where the averaging is taken over the observation period. Clearly,

$$\overline{[\Pi(t)]^2} = \int\int\int_0^L\int dx\, dy\, dx'\, dy' \, [\overline{\Psi(t; x, y)\, \Psi(t; x', y')} + \overline{\varphi(t)\varphi(t)} + 2\,\overline{\Psi(t; x, y)\varphi(t)}].$$

However, due to the independence of the noise and φ the last term vanishes, $\overline{\varphi^2}$ is proportional to the target sound intensity I, and due to the mutual independence of Ψ at one given time at

---

* Here the voltage drop simulating the target sound with the dynamic speaker connected was determined experimentally and taken into account.

two diaphragm points separated by a distance greater than l can be written

$$\overline{\Psi(t; x, y)\, \Psi(t; x', y')} = \overline{\Psi^2} F\left(\sqrt{(x - x')^2 + (y - y^2)}/l\right),$$

where $\overline{\Psi^2}$ is proportional to the noise intensity I_0 and F is a function equal to unity when the argument is zero and equal to zero when the argument is much greater than unity. Consequently, transforming to polar coordinates, we can approximately write [$\rho = \sqrt{(x - x')^2 + (y - y')^2}$]

$$\int F dS \approx \iint \rho\, d\rho\, d\alpha F\,(\rho/l) = l^2 \iint \xi d\xi d\alpha F\,(\xi) = q l^2,$$

where q is a quantity of order unity. Therefore,

$$\iint dS dS'\, F = q l^2 L^2$$

and we have finally

$$\sqrt{\overline{[\Pi(t)]^2}} = \sqrt{(I_0 q l^2 L^2 + I L^4)} \approx \sqrt{q l^2 L^2 I_0}\left(1 + \frac{L^2}{2q l^2}\,\frac{I}{I_0}\right).$$

Inasmuch as $l \sim 1/k$, we can approximately put $(q \sim 1)$

$$L^2/(2q l^2) \sim k^2 L^2/2.$$

   If we attempt to realize the ideal case described above by arraying N microphones within the boundaries of a square, we must clearly do so in such a way that the distance d between the closest microphones will not be greater than l, i.e., $1/k$. Consequently, we must have $kd \leqslant 1$. But $L^2 = Nd^2$ and $N = L^2/d^2 \geqslant k^2 L^2$. Thus, under the best circumstances a dense array of microphones in a square brings us once again to the expression

$$\sqrt{\overline{\Pi^2}} \approx \sqrt{\frac{S^2}{k^2 L^2}\, I_0}\left(1 + \frac{N}{2}\,\frac{I}{I_0}\right),$$

which is obtained if the N microphones are arrayed along a linear baseline with a mutual spacing greater than l. We add that precisely the same result is obtained if the N microphones are arrayed densely along a line rather than in a square area.